Children's Hospital (Boston, Mass.)

Medical and Surgical Report of the Children's Hospital, 1869-1894

Children's Hospital (Boston, Mass.)

Medical and Surgical Report of the Children's Hospital, 1869-1894

ISBN/EAN: 9783337163150

Printed in Europe, USA, Canada, Australia, Japan

Cover: Foto ©berggeist007 / pixelio.de

More available books at **www.hansebooks.com**

MEDICAL AND SURGICAL

REPORT

OF

THE CHILDREN'S HOSPITAL

1869-1894

EDITED BY

T. M. ROTCH, M.D.

AND

HERBERT L. BURRELL, M.D.

BOSTON, U.S.A.

PUBLISHED BY THE BOARD OF MANAGERS

1895

PREFACE.

THE editors have tried in this Report to place the work of THE CHILDREN'S HOSPITAL before the medical public. It consists of Three Divisions: the ADMINISTRATIVE, the MEDICAL, and the SURGICAL.

Dr. Frederick Coggeshall, of Boston, has rendered valuable assistance in reading the manuscript of this Report.

OFFICERS

OF

THE CHILDREN'S HOSPITAL, 1895.

Board of Managers.

PRESIDENT.

OLIVER W. PEABODY . . . . 113 Devonshire Street.

VICE-PRESIDENT.

CHARLES L. YOUNG . . . . 71 Mt. Vernon Street.

TREASURER.

JOSHUA M. SEARS . . . . 206 Sears Building.

SECRETARY.

FRANCIS H. BROWN . . . . 75 Westland Avenue.

MANAGERS.

WILLIAM INGALLS.	H. H. HUNNEWELL.
CHARLES H. FISKE.	G. P. GARDNER.
JOHN G. WETHERELL.	GEORGE THACHER.
JERE. ABBOTT.	F. C. SHATTUCK.
J. P. SPAULDING.	OLIVER AMES, 2D.
E. V. R. THAYER.	CLARENCE J. BLAKE.

Medical Officers.

CONSULTING BOARD.

FRANCIS H. BROWN, M.D.	J. P. OLIVER, M.D.
WILLIAM INGALLS, M.D.	A. M. SUMNER, M.D.
JOHN HOMANS, M.D.	A. T. CABOT, M.D.
ABNER POST, M.D.	H. C. HAVEN, M.D.

CONTENTS.

Division III. — *Surgical.*

PART I.

PART II.

[Diseases of the joint are first considered and grouped together owing to uniformity of treatment ; other diseases are considered in the relative order of their importance.]

𝔇𝔦𝔟𝔦𝔰𝔦𝔬𝔫 III. — *Surgical.*

PART II. *(continued.)*

LIST OF ILLUSTRATIONS.

I. THE ADMINISTRATIVE DIVISION.

THE administrative division contains the history of the Hospital, a description of the buildings, an account of the executive management, a description of the Convalescent Home at Wellesley, Mass., and an account of the out-door relief department of the Hospital.

THE CHILDREN'S HOSPITAL.

THE actual origin of THE CHILDREN'S HOSPITAL, so
far as is shown in writing, dates from the evening
of December 9, 1868, when four gentlemen met at the
residence of the writer of this article to listen to plans
which had formed themselves in his mind, and which
had been talked over for some months with his professional friend, William Ingalls, M. D. The gentlemen
named were Rev. Chandler Robbins, D. D., George H.
Kuhn, J. Huntington Wolcott, and Francis Brown.
Dr. Ingalls and Dr. S. G. Webber were also present,
and aided the writer in explaining the details of the
proposed institution. At a meeting held January 5,
1869, Dr. Robbins, Mr. Kuhn, and Dr. Ingalls, with
Mr. Albert Fearing, Mr. Nathaniel H. Emmons, and
Dr. S. W. Langmaid listened to the plans in further
detail. It was decided to call a meeting of a number of
gentlemen at a still later day; and that gathering was
held, January 23, 1869, at the office of the Massachusetts Hospital Life Insurance Company. Fifteen gentlemen there assembled. The chair was occupied by
Hon. George Tyler Bigelow. It was voted to apply to
the Legislature for authority to form a corporation to
be called THE CHILDREN'S HOSPITAL. The act of incor-

poration was granted by the Legislature, and accepted by the corporators March 22, 1869.

At the very outset of the Hospital its objects were stated to be, —

1. The medical and surgical treatment of the diseases of children;

2. The attainment and diffusion of knowledge regarding the diseases incident to childhood;

3. A system of voluntary nursing, including moral and religious training, by cultivated and experienced Christian women;

4. The training of young women in the duties of nurses.

It is gratifying to realize the fact that the plans which were drawn up have proved adequate to meet all the ends in view, and that, in general, they have been substantially unchanged to the present day.

In all Christian communities the care of children and of the very aged has always called out the deepest interest and the most painstaking devotion of benevolent persons. Especially is the fact recognized that the children of the poorer classes, from their insufficient and poor food, the want of care for their cleanliness, and protection from the weather, their unsanitary abodes, the want of air and sunshine, and, perhaps, equally the want of a healthy moral tone in the family, of kindness and affection, develop a condition of depressed vitality which renders them easily the prey of disease. To this consideration is to be added the thought of every far-seeing philanthropist, that these children are to be future members of society, and that the sound mind of the coming citizen depends largely on a sound body. It is not alone relief from disease which the community

needs; it is the increasing of vitality, and the strengthening of the physical constitution; these can only be
effected by weeks of good food, air and sunshine, which a
wise provision supplies in a well-ordered institution.

Still further, it has been the constant aim of the managers, while endeavoring to cure or alleviate disease,
and to give a tone to their general health, to bring the
patients under the influence of order, purity, and kindness; to introduce them for a season to a brighter side
of life; and by means of intelligent and tender nursing,
by attractive books, pictures, and toys, and by the visits
and attentions of the kind and cultivated, to do something toward beguiling their hours of suffering, quicken-
ing their intellects, refining their manners, and softening
and encouraging their hearts. This object may seem
foreign to the true character of a hospital for the relief
of disease. The managers of The Children's Hospital,
however, hold a different view; and the experience of
twenty-five years assures them that the establishment
and carrying out of these principles have not only been
an important adjunct in the physical care of the children,
but they recall with satisfaction the fact that they have
had an important effect in humanizing and elevating
those under their charge.

Another object which the managers had in view
was to supply a want in our community, which has
been felt in our medical schools; namely, an opportunity to study infantile diseases. These, as every
mother and every nurse knows, are so sudden, fluctu-
ating, and so mysterious, and often so rapid in their
fatality, that they furnish a distinct branch of medical science, the importance of which can hardly be
over-estimated.

The first meeting of the Board of Managers was held March 24, 1869. The Board, as first constituted, was as follows: —

President: NATHANIEL THAYER.
Vice-President: GEORGE TYLER BIGELOW.
Treasurer: JOHN G. WETHERELL.
Secretary: FRANCIS H. BROWN.
Managers: CHANDLER ROBBINS, GEORGE H. KUHN, ALBERT FEARING, NATHANIEL H. EMMONS, CHARLES FAULKNER, ROBERT C. WINTHROP, WILLIAM INGALLS, CHARLES H. FISKE, SAMUEL A. GREEN, ISAAC THACHER, RUSSELL STURGIS, JR., SAMUEL JOHNSON.

At that meeting the Board listened to a paper, signed by Drs. Brown, Ingalls, and Langmaid, suggesting plans for the immediate organization of the hospital, the raising of funds, an appeal to the benevolent members of the community, and other matters of interest. At the meeting of the Board, June 1, 1870, the details were further elaborated so as to include an outline plan of a building for the use of the institution when the success of later years should warrant its construction. In 1881 the present building was erected, substantially on the plan then brought forward.

The suggestions made were at once adopted, with a few changes, and a committee, composed of Messrs. Robbins, Faulkner, Ingalls, and Brown, appointed, with directions to proceed at once to carry on the same, and to put the institution on a working basis. At the same meeting a medical staff was constituted, Drs. William Ingalls and Francis B. Greenough, being made Physicians, and Drs. Francis H. Brown and Samuel W. Langmaid, Surgeons.

At a meeting held March 11, the Committee of Organization reported that they had, as individuals, bought the house No. 9 Rutland Street, which they had found suitable for the purpose, — subject to the approval of the Board. They also stated that they had secured the services, as Lady Superintendent, of Mrs. Adeline Tyler, a lady well known in the community, who, through an already lengthened life, had given her services freely in many good works, and was well fitted, by her experience and personal character, to administer the internal arrangements of the Hospital.

At the same meeting a communication was received from Mrs. J. C. Hooker, announcing that, at the suggestion of the Board, an independent organization of ladies had been formed, chiefly through the agency of Mrs. Robert C. Winthrop and Mrs. Chandler Robbins, the object of which was, under the name of the Ladies' Aid Association, to assist the managers by supplying linen, bedding, dressings, and other needs of the Hospital from time to time. The offer of this association was gratefully accepted, and, at the end of a quarter of a century, it is a pleasant privilege to look back on the work of this adjunct body. While having no control in the work of the Institution, — with unbroken harmony with the Board of Managers, — it has materially strengthened the hands of the authorities by the supply of many articles required for use, by the most generous personal service, by friendly and timely counsel, and by furnishing and carrying on that most important adjunct to the Hospital, the Convalescent Home.

At a meeting of the Board held at the Hospital, No. 9 Rutland Street, on the 19th of July, the Committee of Organization announced that the building was in complete order, and it was then —

"*Voted, that the Hospital be now opened for the reception of patients.*" The first patient was admitted on the next day.

At the meeting of the Board held October 6, at the suggestion of the Medical Staff, that body was enlarged by the addition of two members, and Drs. B. E. Cotting and John Homans were appointed, respectively, Physician and Surgeon.

On the 28th of December, the Board presented to the Corporation its report, with accompanying documents, for the fractional part of the year ending that day, by which it appeared that thirty patients had already received the benefits of the institution, and that fourteen were at that time in the wards. There had been no deaths. At this meeting of the Corporation Mr. Kuhn declined re-election, and Mr. Edward A. Strong was appointed in his place.

In June, 1870, the Hospital had so far outgrown its accommodations that the Board began to consider plans to secure new quarters, and on the 24th of the same month the Board authorized a committee to close negotiations with Mr. E. C. Bailey, the owner of the estate No. 1429 Washington Street, for the use of that property for a series of years. The estate on Rutland Street was sold in May, 1871.

In October, 1876, the lease of the house was renewed for three years.

The new building was occupied July 28, 1870, and was held until the removal to the Hospital on Huntington Avenue.

On the 7th of November, 1871, Mr. Sturgis resigned, and his place was filled at the next meeting of the Corporation.

On the 12th of March, 1872, Mrs. Tyler presented her resignation of the office of Lady Superintendent, and Sister Theresa, of the Protestant Episcopal Sisterhood of St. Margaret, was chosen to fill her place. Mrs. Tyler died in January, 1875.

Dr. B. E. Cotting resigned his position of Physician October 1, 1872, and Dr. W. L. Richardson was chosen to fill his place.

At the meeting of March 4, 1873, a communication was received from Mr. E. R. Mudge, of Emanuel Church, in reference to some plan for the establishment of a Training School for Nurses, which was duly referred to a committee.

In October, 1873, Dr. Frank W. Draper was made an assistant to the Medical Staff.

At the Corporation meeting of December 28, 1873, Dr. A. M. Sumner was elected Physician, in place of Dr. Greenough, who declined re-election.

March 3, 1874, the Board considered the establishment of a Sanitarium, or Convalescent Home, in the country, to be occupied during the summer months. Accommodations for six children, a lady, and a nurse were provided in the town of Weston, from the first day of July, and were occupied until the early days of the autumn. Dr. John Smithwick, of Weston, gave his services gratuitously during the season. Nineteen children were cared for during July and August, at an expense of $221.97.

October 6, 1874, Dr. W. L. Richardson resigned the position of Physician. Dr. Draper was promoted to Physician, and Dr. F. G. Morrill was made Assistant.

December 1, 1874, the Board established a department for out-patients, and the same was carried on in a room at 1725 Washington Street by members of the Medical Staff. The expenses of furnishing the room, and of carrying it on for a year, were met by Dr. Sumner, of the Staff. At the close of the year the Department was removed to the Hospital.

January 5, 1875, a communication was received from the Ladies' Aid Association concerning the foundation of a permanent Convalescent Home, and a Committee of Conference with the Association was appointed. The proposition made by the ladies was accepted at the meeting of February, and the Institution was opened, in a leased house in Wellesley, May 31; and continued in the same house until the purchase, by the Ladies' Aid Association, of another estate in the same town in June, 1877. During the entire establishment of the Home to the present day its expenses have been borne by the Association. Thirty-nine children were cared for during the first summer.

Dr. L. R. Stone, of Newton, gave his services to the Home during two years.

Mr. Albert Fearing, a member of the Board of Managers from the opening of the Hospital, died May 24, 1875; and at the meeting held June 1 his death was appropriately noticed. At the meeting in October, Mr. Alanson Bigelow was chosen to fill his place.

At the same meeting Dr. Draper resigned his position as Physician. Dr. Morrill was advanced to his place, and Dr. Thomas Waterman was chosen Assistant.

November 7, 1876, Mr. W. P. Kuhn was chosen Treasurer, to serve during the absence of Mr. Wetherell.

December 5, 1876, Dr. Homans presented his resignation as Surgeon. Dr. Ingalls was made Surgeon in his place, and Dr. Joseph P. Oliver, Physician, in place of Dr. Ingalls.

In January, 1878, Dr. Brown declined re-election as Surgeon, and Dr. E. H. Bradford was chosen in his place; Dr. E. G. Cutler was made Pathologist, and Dr. F. H. Davenport, Assistant Physician.

In April, 1878, Mr. N. II. Emmons, a manager from the opening of the Hospital, died.

May 6, 1878, the Medical Staff presented a communication to the Board, asking that a Consulting Board be constituted, and suggested the names of Drs. Brown and Homans for positions on such board. The suggestion was adopted by the Board. At the same meeting Mr. Mudge resigned his position on the Board; and at the meeting of the Board in January, 1879, Mr. H. G. Pickering was elected to his place. Dr. U. O. B. Wingate was elected Physician to the Convalescent Home.

Sister Frances, who had served the Hospital as Lady Superintendent since September 4, 1877, during Sister Theresa's absence in Europe, was relieved by the return of the latter, October 7, 1879.

December 2, 1879, on motion of Mr. Pickering, a Committee was constituted having for its object the consideration of plans for a new and permanent hospital. This Committee was composed of Messrs. Robbins, Thacher, Wetherell, Pickering, and Brown. January 6, 1880, the Committee reported that it seemed best to proceed at once to obtain means to erect a permanent building for the use of the Hospital, capable of containing one hundred beds. The subject was fully discussed at this meeting, and at many subsequent ones; assistance was asked from architects, from the authorities of other hospitals, and others.

Messrs. Faulkner, Thacher, and Brown were consti-

tuted a Building Committee, June 28, 1880, and were
authorized, July 2, to purchase the site on Huntington
Avenue, corner of Camden (later Gainsborough) Street,
containing 31,075 feet. Plans were drawn for the new
building by Messrs. Winslow and Wetherell, architects,
with Mr. Nathaniel J. Bradlee as consulting architect.

March 4, 1881, the Building Committee were directed
to erect the central portion and the west wing of the
new Hospital, in conformity with the plans offered.
Mr. Phillips was appointed to serve on the Committee
in place of Mr. Thacher. The building was erected by
Messrs. Coburn and Root, masons, and Morton and
Chesley, carpenters.

Work was begun April 26, 1881, the corner-stone
was laid June 13 of the same year, and the building was
dedicated to its benevolent uses December 26, 1882.

The building occupied by the Hospital consists of
a central or administrative building, with wings on either
side containing wards, and an adjacent building for the
use of the out-patient departments. The entire construction is of brick, in the Renaissance style of architecture,
thoroughly and substantially built of the best material.
The entire building is 318 feet long; the centre building, 44, the wings, each 113, and the out-patient building,
48. The centre or administrative building contains, on
the first floor, the managers' and staff room, reception
rooms, sitting-rooms and dining-rooms for the house
officers and Sisters, an operating room, an etherizing
room, a dispensary, and a small ward, etc. The second
floor of this building is devoted to the Sisters who
have charge of the work. The third floor has five
rooms for private patients, and the fourth floor accom-
modates the nurses.

ROOM OF THE MANAGERS AND STAFF.

The west wing is devoted to the surgical patients, who are disposed of in two wards, of twenty beds each. The beds are placed along the sides of the ward.

The east wing is mostly set apart for medical cases. The lower floor has two large wards, for boys and girls respectively. The second floor is made into small wards

MEDICAL WARD.

and single rooms, for the care of such cases as it may be well to treat separately.

Over these wards the space has been so utilized that a play-room is secured: it occupies nearly the entire space over this wing. This provision has been found a valuable adjunct in the care of patients, giving a change of air and scene as well as amusement, — essential aids to the improvement of patients in the wards.

The building for the out-patient department was especially erected for the purpose. Its first floor has a large waiting-room for surgical out-patients, with numerous small administrative rooms. From this waiting-room extend a series of small rooms where children are prepared for observation, and apparatus is fitted. There is

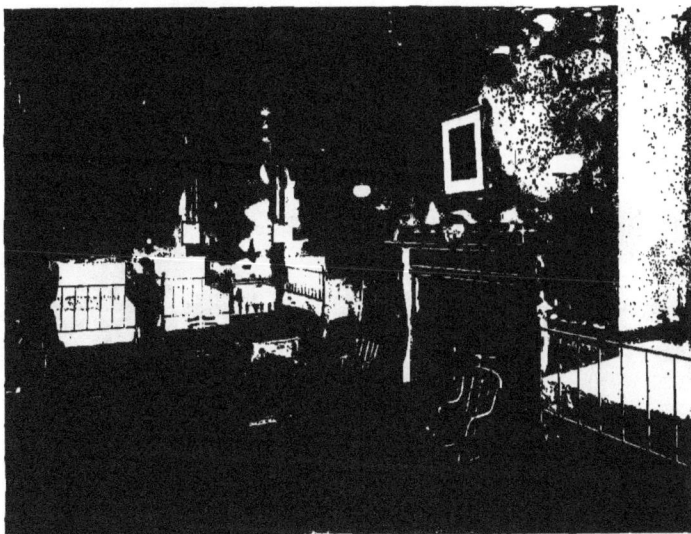

SMALL MEDICAL WARD.

an operating-room, and a room for the application of massage and gymnastic exercise.

The second floor of this building is used for medical patients, as well as for those with diseases of the nose and throat, and for nervous diseases. A room for photography and a lecture-room complete the rooms on this floor. The upper floor, which is reached by a separate entrance, is devoted to such contagious cases as occur in the Hospital.

The basement of this building is divided into rooms used for a large museum, a plaster-of-Paris work-room, a workshop for house officers, and a surgical-appliance shop.

The Hospital is warmed by forcing heated air directly into the wards and rooms, by direct radiators in the rooms themselves, and by open fires.

THE LECTURE ROOM.

Ventilation is secured by shafts starting near the floor and at the ceiling of the rooms, by ridge ventilation, open fireplaces, and by double windows and direct ventilation. As the building lies with its whole length to the southwest, it receives a full supply of sunlight each day.

The walls are painted with three coats of paint and varnish; most of the floors are of hard pine, underlaid with cement, the floors of bath-rooms and water-closets are of marble. The plumbing is largely of the open character. The water-closets are cut off from any communication with the wards.

December 28, 1881, it was voted by the Corporation that any person giving at one time not less than five hundred dollars should have the privilege of affixing the name of a child to a bed in the Hospital, subject to the decision of the Board in each case. It was understood by this vote, however, that this permission did not carry with it any right or control over the bed named. Up to the present time the following memorial beds have been named, viz. : —

SULLIVAN AMORY, 1882.
LIVINGSTON WADSWORTH, 1882.
DOROTHY QUINCY, 1883.
GEORGE F. KIMBALL, 1884.
LOUISE JULIA TUCKER, 1885.
SARAH KITTREDGE GOODNOW, window and two beds, 1886.
ISAAC THACHER.
SARAH ELIZA THACHER.
MARY ELLEN PUTNAM.
RACHEL TOWER TARBELL.
STEPHEN WHEATLAND.

July 5, 1882, on the recommendation of the Medical Staff, the Board decided to appoint as house officers, an Interne and an Externe, to assist the Medical officers in the care of the patients.

The first meeting of the Board in the new building was held November 7, 1882.

April 1, 1884, a communication was received from the Medical Staff relating to the construction of a small building in the hospital yard for the purposes of a workshop. Authority to erect such a building was conferred June 3.

April 6, 1886, the Board considered the imperative demand of the out-patient department for larger accommodations, and the importance of the work which it was accomplishing. At the next meeting, Mr. George Thacher, one of the Board, offered to give the sum of two thousand dollars toward the erection of a building for the out-patient department, as a memorial of his father, Mr. Isaac Thacher, a former member of the Board, and of his sister, Miss Sarah E. Thacher, both lately deceased. February 1, 1887, the Board authorized the erection of a building, for an out-patient department, on the land, about 8000 feet, recently acquired. The building was completed, and put to its appropriate use, October 1, 1888.

In May and June, 1889, the Board considered plans for the erection of the east wing of the Hospital. It was completed in due season. It was then thought best to reorganize the assignment of the wards in the Hospital, so that the west wing might be exclusively occupied by the surgical cases, and a large part of the east wing by the medical. In addition, there was secured, over a considerable part of the east wing, the commodious play-room, which has been already described. Rooms for the house officers, store-rooms, an elevator, and other conveniences were supplied.

At a meeting of the Board held February 7, 1893, a Committee was appointed to prepare and despatch to Chicago an exhibit, comprising full sets of the plans of the Hospital and Convalescent Home, a description of the working of the various departments, reports, photographs of the buildings, wards, surgical appliances, patients, etc., for display at the Columbian Exposition. Such a plan was fully carried out, and the Hospital was

THE PLAY ROOM.

worthily represented at the Exhibition. The Board are assured that the exhibit was consulted by many persons interested in the care of children, and they trust that it accomplished a good work.

The publication of this volume was suggested to the Board by the Medical Staff, and was authorized at the meeting held May 5, 1891. It was suggested, with the view of presenting to the community, first, the history of an institution organized and maintained for the care of sick and disabled children; secondly, in order to show, by the contributions of the members of its medical staff, how much good work had been done, and the most efficient methods known to medical science of accomplishing its object; and, thirdly, of offering carefully prepared articles, based largely on the practice in the Hospital, on various topics of importance in the science of medicine.

FRANCIS H. BROWN, M. D.

A LIST OF THE OFFICERS

AND OTHERS CONNECTED WITH THE HOSPITAL FROM ITS ORGANIZATION.

OFFICERS.

Presidents.

NATHANIEL THAYER, Mar. 22, 1869. Died in office, Mar. 8, 1883.
ROBERT C. WINTHROP, Dec. 28, 1883. Died in office, Nov. 16, 1894.
OLIVER W. PEABODY, Dec. 28, 1894 —

Vice-Presidents.

GEORGE TYLER BIGELOW, Mar. 22, 1869. Died in office, April 12, 1878.
ROBERT C. WINTHROP, Dec. 28, 1877 — Dec. 28, 1883.
JOHN C. PHILLIPS, Dec. 28, 1883. Died in office, Mar. 1, 1885.
OLIVER W. PEABODY, Dec. 28, 1885 — Dec. 28, 1894.
CHARLES L. YOUNG, Dec. 28, 1894 —

Treasurers.

JOHN G. WETHERELL, Mar. 22, 1869 — Oct. 24, 1876.
WILLIAM PUTNAM KUHN, Oct. 24, 1876 — Nov. 7, 1877.
JOHN G. WETHERELL, Dec. 28, 1877 — Dec. 28, 1886.
JOSHUA M. SEARS, Dec. 28, 1886 —

Secretary.

FRANCIS H. BROWN, M.D., Mar. 22, 1869 —

Managers.

CHANDLER ROBBINS, Mar. 22, 1869. Died in office, Sept. 11, 1882.
ALBERT FEARING, Mar. 22, 1869. Died in office, May 24, 1875.
GEORGE H. KUHN, Mar. 22, 1869 — Dec. 28, 1869. Died Feb. 21, 1879.
NATHANIEL H. EMMONS, Mar. 22, 1869. Died in office, April, 1878.
CHARLES FAULKNER, Mar. 22, 1869. Died in office, April 1, 1885.
ROBERT C. WINTHROP, Mar. 22, 1869 — Dec. 28, 1877.
WILLIAM INGALLS, M.D., Mar. 22, 1869 —
CHARLES H. FISKE, Mar. 22, 1869 —

SAMUEL A. GREEN, M.D., Mar. 22, 1869 — Dec. 28, 1883.
ISAAC THACHER, Mar. 22, 1869. Died in office, Feb. 5, 1883.
RUSSELL STURGIS, Jr., Mar. 22, 1869 — Dec. 28, 1871.
SAMUEL JOHNSON, Jr., Mar. 22, 1869 — Dec. 28, 1871.
EDWARD A. STRONG, Dec. 28, 1869 — Dec. 28, 1873.
JERE ABBOTT, Dec. 28, 1871 —
GEORGE D. HOWE, Dec. 28, 1871 — Dec. 2, 1880.
ENOCH R. MUDGE, Dec. 29, 1873 — May 6, 1878. Died 1881.
ALANSON BIGELOW, Dec. 28, 1875. Died in office, Feb. 29, 1884.
W. P. KUHN, Dec. 28, 1877 — Dec. 28, 1880.
JOHN D. SPAULDING, Dec. 28, 1878 —
HENRY G. PICKERING. Dec. 28, 1878 — March 2, 1886.
J. M. MANNING, Dec. 28, 1880 — June 6, 1882. Died 1882.
JOHN C. PHILLIPS, Dec. 28, 1880 — Dec. 28, 1883.
ROLAND C. LINCOLN, Dec. 28, 1882 — Feb. 3, 1885.
JOSHUA M. SEARS, Dec. 28, 1882 — Dec. 28, 1886.
O. W. PEABODY, Dec. 28, 1883 — Dec. 28, 1885.
PHILLIPS BROOKS, Dec. 2, 1883 — Nov. 3, 1885. Died 1893.
CHARLES L. YOUNG, Dec. 28, 1883 — Dec. 28, 1894.
E. V. R. THAYER, April 1, 1884 —
FREDERIC L. AMES, Dec. 28, 1885. Died in office. Sept. 12, 1893.
H. H. HUNNEWELL, Oct. 6. 1885 —
GEORGE P. GARDNER, Dec. 1, 1885 —
GEORGE THACHER, Dec. 28, 1885 —
JOHN G. WETHERELL, Dec. 28, 1886 —
CHARLES H. JOY, May 4, 1886 — Dec. 28, 1887. Died
FREDERICK C. SHATTUCK, M.D., Dec. 28, 1887 —
OLIVER AMES, 2nd, Dec. 5, 1893 —
CLARENCE J. BLAKE, M.D., Dec. 28, 1894 —

MEDICAL OFFICERS.

Consulting Board.

FRANCIS H. BROWN, 1878 —
JOHN HOMANS, 1878 —
WILLIAM INGALLS, 1881 —
JOSEPH P. OLIVER, 1884 —
ALLEN M. SUMNER, 1887 —
ARTHUR T. CABOT. 1890 —
ABNER POST, 1890 —
H. C. HAVEN. 1892 —

Physicians.

WILLIAM INGALLS, Mar. 24, 1869 — Dec. 5. 1876.
FRANCIS B. GREENOUGH, Mar. 24, 1869 — Dec. 28, 1873.
BENJAMIN E. COTTING, Oct. 6, 1869 — Oct. 1, 1872.
WILLIAM L. RICHARDSON, Oct. 1, 1872 — Oct. 6, 1874.

ALLEN M. SUMNER, Jan. 6, 1874 — Dec. 28, 1877.
FRANK W. DRAPER, Oct. 6, 1874 — Oct. 5, 1875.
F. GORDON MORRILL, Oct. 5, 1875 —
JOSEPH P. OLIVER, Dec. 5, 1876 — July 1, 1884.
H. C. HAVEN, Aug. 5, 1884 — Feb. 2, 1892.
THOMAS M. ROTCH, Feb. 2, 1886 —
EDWARD M. BUCKINGHAM, Mar. 1, 1892 —

Surgeons.

FRANCIS H. BROWN, Mar. 24, 1869 — Jan. 1, 1878.
SAMUEL W. LANGMAID, Mar. 24, 1869 — Jan. 5, 1886.
JOHN HOMANS, Oct. 6, 1869 — Dec. 5, 1876.
WILLIAM INGALLS, Dec. 5, 1876 — July 5, 1881.
EDWARD H. BRADFORD, Jan. 1, 1878 —
ARTHUR T. CABOT, July 5, 1881 — Dec. 3, 1889.
ABNER POST, Mar. 2, 1886 — Dec. 6, 1887.
HERBERT L. BURRELL, Jan. 4, 1888 —

Surgeon for Diseases of the Throat.

SAMUEL W. LANGMAID, Jan. 5, 1886 —

Ophthalmic Surgeon.

OLIVER F. WADSWORTH, April 1, 1890 —

Aural Surgeons.

CLARENCE J. BLAKE, April 1, 1890 — June 5, 1894.
HENRY L. MORSE, June 5, 1894 —

Assistants to the Medical Staff.

FRANK W. DRAPER, Oct. 7, 1873 — 1874.
F. GORDON MORRILL, Oct. 6, 1874 — Oct 5, 1875.
THOMAS WATERMAN, Oct. 5, 1875 — Jan. 1, 1878.
FRANK H. DAVENPORT, Jan. 1, 1878 — Jan. 6, 1880.

Assistant Physicians.

FRANK H. DAVENPORT, Jan. 6, 1880 — July 5, 1881.
H. C. HAVEN, July 5, 1881 — Aug. 5, 1884.
T. M. ROTCH, Nov. 7, 1882 — Feb. 2, 1886.
EDWARD M. BUCKINGHAM, Aug. 5, 1884 — Mar. 1, 1892.
C. W. TOWNSEND, Feb. 5, 1895 —
A. H. WENTWORTH, Feb. 5, 1895 —
G. A. CRAIGIN, Feb. 5, 1895 —

Assistant Surgeons.

ARTHUR T. CABOT, Dec. 29, 1879 — July 5, 1881.
ABNER POST, July 5, 1881 — Mar. 2, 1886.

HERBERT L. BURRELL, Mar. 2, 1886 — Jan. 4, 1888.
H. W. CUSHING, Jan. 4, 1888 —
R. W. LOVETT, Dec. 28, 1893 —
E. G. BRACKETT, Dec. 28, 1893 —
J. E. GOLDTHWAIT, Dec. 28, 1893 —

Assistants in the Medical Out-Patient Department.

THOMAS F. SHERMAN, Feb. 7, 1888 — Dec. 4, 1888.
CHARLES W. TOWNSEND, Feb. 7, 1888 — Dec. 4, 1888.

Assistants in the Surgical Out-Patient Department.

FRANCIS S. WATSON, Dec. 4, 1883 — Mar. 3, 1885.
HERBERT L. BURRELL, Dec. 4, 1883 — Mar. 2, 1886.
HAYWARD W. CUSHING, Mar. 2, 1886 — Jan. 4, 1888.
ROBERT W. LOVETT, Nov. 1, 1887 — Dec. 4, 1888.

Assistant Physicians to Out-Patients.

THOMAS F. SHERMAN, Dec. 4, 1888.   Died in office, Sept. 26, 1893.
CHARLES W. TOWNSEND, Dec. 4, 1888 — Feb. 5, 1895.
A. B. WENTWORTH, April 5, 1892 — Feb. 5, 1895.
GEORGE A. CRAIGIN, Oct. 3, 1893 — Feb. 5. 1895.

Assistant Surgeons to Out-Patients.

ROBERT W. LOVETT, Feb. 7, 1888 — Dec. 28, 1893.
EDWARD G. BRACKETT, Dec. 4, 1888 — Dec. 28, 1893.

Assistant in Diseases of the Throat.

ALGERNON COOLIDGE, Jr., April 1, 1890 —

Assistant Aural Surgeon.

H. L. MORSE, April 1, 1890 — June 5, 1894.

Neurologist and Electrician.

WILLIAM N. BULLARD, Jan. 4, 1888 —

Pathologist.

ELBRIDGE G. CUTLER, Jan. 1, 1878 — Dec. 29, 1879.
WILLIAM F. WHITNEY, Dec. 29, 1879 —

Physicians to the Convalescent Home.

JOHN SMITHWICK, 1874.   U. O. B. WINGATE, 1878.
LINCOLN R. STONE, 1875.   G. J. TOWNSEND, 1883.

LADY SUPERINTENDENTS.

Mrs. Adeline Tyler, 1869 — 1872.
Sister Theresa, 1872 — 1877.
Sister Frances, 1877 — 1879.
Sister Theresa, 1879 — 1888.
Sister Maria, 1888. Died in office,
  Dec. 16, 1892.

Sister Anna, Dec. 16, 1892 — Jan.
  3, 1893.
Sister Caroline, Jan. 3, 1893 —

HOUSE OFFICERS.

Henry S. Otis, 1882.
George Haven, 1882.
W. F. Knowles, Jr., 1882.
George F. Tucker, 1883.
Clarence W. Spring, 1884.
John W. Perkins, 1884.
William B. Fiske, 1885.
Homer Gage, 1885.
Percival J. Eaton, 1885.
Charles L. Scudder, 1886.
Nathaniel S. Hunting, 1886.
William E. Fay, 1887.
Rufus E. Darrah, 1887.
Joel E. Goldthwait, 1888.
Frank E. Peckham, 1888.
John H. Huddleston, 1889.

Edward H. Nichols, 1889.
Harold G. Gross, 1890.
George W. Fitz, 1890.
Frank S. Whittemore, 1890.
Frank A. Higgins, 1891.
Edwin P. Stickney, 1891.
Rupert Norton, 1891.
Julio Selva, 1892.
Frederick H. Baker, 1892.
William Cogswell, Jr., 1892.
Calvin Gates Page, 1893.
Harrison D. Jenks, 1893.
Charles F. Painter.
Herbert J. Hall.
Eugene C. Wylie.
Frederic A. Washburn, Jr.

GRADUATES OF THE TRAINING SCHOOL FOR NURSES.

Katharine E. Ambrose.
Jane H. Wetmore.
Elizabeth M. H. Wetmore.
Ida C. Smith.
Nina M. Brown.
Helena S. de Veer.
Maria K. Watt.
Minnie E. Travis.
Jennie Gray.
Emilie M. Spaul.

Annie Spaul.
Sibyl M. Holbrook.
Susan B. Johnson.
Edith M. Tuttle.
Gertrude H. Tuttle.
Clara W. Barstow.
Nevah E. Sanborn.
Mary R. Roberts.
Susan A. Gregory.

THE EXECUTIVE MANAGEMENT OF THE HOSPITAL.

THE Board of Managers place the immediate care of the hospital with the Sisters of St. Margaret, one of whom is annually appointed Lady Superintendent. Under her are others of the same body, who are in charge of the medical and surgical nursing.

The Superintendent admits all patients, orders all supplies, keeps a daily record of the number of children in the wards, answers all letters of inquiry about the hospital and Training School, keeps a record of and acknowledges all donations sent to the hospital, receives all visitors, and shows them over the building, makes a daily visit of inspection to every part of the building, engages and dismisses all nurses and employees, and sends in a monthly report, [*Form A.*] also a list of expenditures, to the Board of Managers.

The Sisters in charge of the wards make all visits with the physician or surgeon, see that all orders are promptly and intelligently carried out, and have the general charge of the nurses in the wards.

The Housekeeper has charge, under the Superintendent, of the kitchen and laundry, orders daily all provisions, groceries, etc., and has the immediate supervision of all the servants.

The Visiting Day for parents is Wednesday, from eleven to twelve o'clock ; Sundays from three to four P.M. (for fathers only).

Form A.

MONTHLY REPORT OF THE SUPERINTENDENT.

To the Visiting Committee: —
The undersigned makes the following report for the month ending
................................*189* .

Number of Patients remaining in Hospital by
    last report. .   . .   . . .
    Admitted . .
    Discharged . .
    Remaining this day . . . . . .
Number of Patients remaining at Convalescent
    Home by last report   . . . .
    Admitted . . .   .
    Discharged . .
    Remaining this day . . . . .
Number of Patients at Hospital and Convales-
    cent Home . . . . . .

    In the case of those discharged, the result has been as follows: —
    Well . . . . . .
    Relieved . . .
    Not relieved . . .
    Not treated . .
    Died. . . . .
        Cause of death ....

Total number of patients admitted to this date . . .
        Respectfully submitted,
                 ....
                           Superintendent.

Recently it has been found necessary to exclude the parents from the wards in the hospital. Coming as they do from unknown places, from which they may bring contagious diseases, they are a menace to the children in the hospital. The children are brought out on a wheeled carriage, when necessary, to the parents, who remain in a room provided for the purpose.

ROUTINE WORK IN THE WARDS.

The nurses go on duty at 6.45 A. M. The temperatures of the patients are first taken, and very sick patients attended to. The children's breakfast is served at 7 A. M. In each ward of twenty beds there are two nurses and a ward-maid; the latter does all the sweeping, dusting, and cleaning. After breakfast is served, each nurse, to whom has been assigned the care of ten beds, starts on her own side at No. 1 or No. 20 bed. She is required to bathe each child, and, as she makes the beds, to turn the mattresses, etc. The ward is in order at 10 A. M., when all the children are given a glass of milk, after which until twelve any extra work is done (new extensions applied, etc.). A written requisition is sent each day to the drug-room for any medicines which may be needed in the wards; also to the operating-room nurse for dressings, bandages, etc., who also keeps on hand frame covers, plaster, and stocking extensions, perineal straps, etc.

The children's dinner is served at 12 M.; after which the ward is swept, the children washed, the beds are made tidy, etc. During the afternoon there is always extra work to be done, new patients to be bathed, the clean linen to be put away, patients to be prepared for operation the next day, etc. At 4 P.M., all temperatures are taken and recorded, and the beds turned down.

At 5 P.M. the children have supper, after which the convalescent children are put to bed, the backs of all bed patients rubbed with alcohol, the crumbs brushed out of the beds, and the children made comfortable for the night. The wards are closed at 6 P.M.

A responsible person is always in charge of the wards, the maids remaining with the children while the nurses are absent for class, lectures, or meals.

The following diets are in use in the hospital : —

DIET.

MILK DIET. — Milk only.
SEMI-SOLID. — Pudding, milk, broth, bread.

HOUSE DIET.

BREAKFAST. — Milk, bread and butter, crackers, oatmeal.
DINNER. — Roast beef or mutton, boiled rice, potatoes, pudding, broth.
SUPPER. — Milk, bread and butter, crackers.

The Training School for nurses at The Children's Hospital was organized in 1890. The nurses are given a two years' course of practical and theoretical training in the diseases of children. Young women between the ages of twenty and thirty are received. Application for admission is made to the Superintendent, and, if there is no apparent objection, they are given an application-blank. This is filled out by the applicant herself, and is as follows : —

TRAINING SCHOOL FOR NURSES.

APPLICATION FOR ADMISSION.

Name (in full)..

Present address...

Your present occupation or employment...

Place and date of birth..

In what schools were you educated?...

Are you strong and healthy, and have you always been so?.........................

Have you any physical defects?...

Are you free from any domestic responsibility, so that you are not liable to be called away?..

Give the names and addresses of two persons, not relatives, for reference.

....

Having read and clearly understood the above questions, I declare the answers to them to be correct; and if accepted as a candidate, I will agree to remain in the School for two years, and will in all respects conform with the rules of the Hospital. It is understood that the officers of the Hospital reserve the right to terminate the connection of any nurse with the School for reasons which they may deem sufficient.

(Signed)...

Boston, ...*189 .*

Examined and approved.

....
Lady Superintendent.

Approved and admitted after one month's probation.

....
Chairman Executive Committee.

Date.................................

If the applicant is satisfactory, she is placed on the waiting-list, and is received as a probationer for one month. If accepted at the end of that time, she is given three uniforms, caps, and aprons, and signs an agreement to remain at the Training School two years and conform to the rules of the hospital. About one-third of the applicants are accepted. The nurses are paid, —

$8.00 per month for the first six months.
13.00 "   "   "   " second six "
15.00 "   "   "   "   " year.

The nurses' term of service is divided as follows: —

Eight months in the Surgical Wards.
Eight   "   "   " Medical   "
Three   "   on night-duty.
Two   "   in the Operating Room.
Two   "   "   " Out-Patient Department.
One month in the Convalescent Home.

The day-nurses' working hours are from 6.45 A. M. to 6 P. M. They are given an afternoon off duty every week; every other Sunday from 10 A. M. to 2 P. M., and one hour off every afternoon; also one hour for lecture on Thursday, and one hour for class on Saturday. Seniors: lecture, Thursday, 4 to 5 P. M.; class, Saturday, 2 to 3 P. M. Juniors: lecture, Tuesday, 4 to 5 P. M.; class, Friday, 2 to 3 P. M.

The night-nurses' hours on duty are from 8 P. M. to 7 A. M.

They are given class-work each week, by one of the Sisters, in the practical details of nursing. The text-books used are, —

Anatomy and Physiology, Kimber.
Materia Medica, Dock.
Text-Book of Nursing, Clara Weeks.

Each week, from October to July, they are given a lecture, by one of the Staff, on the following subjects:

FIRST YEAR.

LECTURES AND DEMONSTRATIONS.

1. Muscular grouping. Vital organs.
2. Topographical Anatomy.
3. Heart and Lungs. Circulation. Pulse. Respiration.
4. General Hygiene. Secretion. Excretion. Care of the Skin.
5. Physiology of digestion.
6. Preparation of food.
7. Serving of food.
8. Nutritive enemata.
9. General nursing. Baths.
10. Clothing. Care of sick-room, ventilation, temperature, etc.
11. Use of disinfectants and deodorizers.
12. Methods of disinfecting a room, bed-clothing, and individual.
13. Medicines. Preparation.
14. Medicines. Administration.
15. Poisons.
16. Antidotes.
17. Etherization.
18. Preparation of patient for operation.
19. Care of patient during and after operation.
20. Shock, collapse, and hemorrhage.
21. Theory of wounds, including inflammation.
22. Suppuration, Septicaemia. Pyaemia. Gangrene, Erysipelas.
23. Antisepsis.
24. Disinfectants. Deodorizers.
25. Nursing in skin diseases.

SECOND YEAR.

LECTURES AND DEMONSTRATIONS.

1. Surgical dressings.
2. Antiseptic. Open dressings.
3. Burns, blisters. cupping, leeches.
4. Special nursing in fevers.
5. Condition of joints.
6. Mechanism of joints.
7. Subjective symptoms in joint diseases.
8. Method of handling patients with joint disease.

9. Lecture on care of orthopedic cases. Demonstration of apparatus. Use and object of apparatus. Method of application. Faulty application. In cases of spinal disease.
10. Spine. Traction and application of extension.
11. Demonstration.
12. Hip. Traction and application of extension.
13. Demonstration.
14. Knee.
15. Demonstration of care of clubfeet.
16. Demonstration in use of plaster of Paris.
17. "   "   "   "   "   "   "
18. Fractures and Dislocations.
19. Demonstration of bandages. Demonstration of splints.
20. Massage.
21. "
22. "
23. Special nursing in nervous diseases.
24. Contagious diseases. Tracheotomy and Intubation.
25. Diseases of naso-pharynx and mouth.
26. Eye.
27. Ear.
28. Autopsies and care of the dead.
29. Duties and conduct of nurses in private nursing.

They are also given practical training on —

The making and changing of beds.
The bathing of the patients.
The application and care of splints and braces.
The application of plaster-of-Paris bandages.
The care of children on the bed-frame.
The administration of medicine.
The giving of enemata.
The care and ventilation of the sick-room.
The care of the operating-room.
The preparation of aseptic dressings.
The dressing of wounds.
The care of skin-diseases.

They are also taught that the first principle in the care of children is to keep them happy and occupied.

If they pass the required examination at the end of two years, they are given a diploma.

SISTER CAROLINE.

THE CONVALESCENT HOME OF THE
CHILDREN'S HOSPITAL.

A T the time of the foundation of The Children's Hospital, in Boston, in 1869, an independent organization of ladies was formed, the object of which was, under the name of the "Ladies' Aid Association," to assist the Managers by supplying linen, bedding, dressings, and other needs of the hospital, from time to time. From the day of its organization to the present day, these expenses have been borne by the Ladies' Aid Association, incorporated in 1885, under the name of the " Convalescent Home of The Children's Hospital."

In 1874, a small house in the town of Weston was occupied as a Convalescent Home during the summer months, with accommodation for six children, a lady as Superintendent, and a nurse. Nineteen children were received during July and August.

In 1875, the Ladies' Aid Association proposed the foundation of a permanent Convalescent Home, and, in a conference with the Managers of the hospital, their plan was accepted, and a house in Wellesley was hired, and occupied by them until 1877, when, through the generosity of Miss Caroline A. Brewer, and many other kind friends, an estate in the same town was bought, and occupied until 1890.

In June, 1890, it became necessary for the Managers
of the Convalescent Home to make plans for the future
accommodation of the children, as the increased size
of the hospital rendered the Home quite inadequate.
It was decided to buy for eight thousand dollars an
estate of thirty-three acres, in Wellesley Hills. This
would allow proper space for a large building with
ample grounds. The land became a gift to the Corporation, owing to the generosity of H. H. Hunnewell, Esq.
Plans were drawn by Messrs. Shaw & Hunnewell, architects, for a new and permanent building, the cost of
which was estimated to be forty-eight thousand dollars.

The friends of the hospital, and the public generally,
were asked for aid, and the response was hearty and
liberal. The Building Committee was directed to go
on with the erection of the Home, and on June 16,
1892, the building was opened and dedicated.

The memory of that dedication will always be united
in the minds of the men and women who were present,
with the date of the death of the beloved President of
the Convalescent Home, Mrs. Robert C. Winthrop. Of
the many benevolent interests which engaged the love
and labor of Mrs. Winthrop, the Convalescent Home
was the one nearest to her heart. From the organization of the Society, in 1869, to her death, she was its
President. The completion of her work was the new
Home. By her unwearied efforts the funds were raised,
and she watched closely the progress of the building
until it was ready for the children. Her last days were
spent in preparing for its dedication and opening; but
she died of a sudden illness before the dedication day
ended. All who worked with her for this especial good
work will ever testify to her faithful service.

THE CONVALESCENT HOME OF THE CHILDREN'S HOSPITAL. — WELLESLEY.

As one can see by the accompanying plan, the Convalescent Home is a square building of three stories, with two wings, one on each side, — each wing having windows on three sides. The whole building, not including piazzas, covers 10,221 feet of land.

The entrance-hall has in the centre an opening to the second story, giving light and supplying sufficient general circulation of air. At the right of the entrance is the dining-room for the Superintendent and assistants; on the left is the business-office or reception-room. Behind this dining-room is that of the children, having a large bay-window in it, looking west. Back of the reception-room is a large play-room, measuring 28 x 39 feet, with double fire-place; at the end of this room, toward the south, is a piazza without roof, that no sunshine be shut out. To the right and left of the play-room are two small piazzas connected with the play-room, in sheltered positions formed by the angles of the building, which can be covered in with glass for cold weather. On this floor, occupying intervening space, are two rooms for nurses, the pantry, and store-closets.

The wings of the building contain the wards, each of which covers 1869 feet and holds 23 beds. Both wards are provided with double fire-places, in addition to the means for the general heating of the building. Bathrooms, store-rooms, a bandage-room, adjoin the wards.

In the second story are the rooms of the Superintendent and assistants. There is also in this story an isolated ward, for use in case of the outbreak of any contagious disease. The third story is used for the rooms of those employed in the household care of the Institution. The Home is under the charge of the Sisters of St. Margaret, who for so many years have had the superintendence of the hospital. The children are sent from the wards of the hospital, or the "out-patient department." Each Tuesday afternoon brings new faces, and takes away old friends. If the new arrivals have come from the hospital wards, they are at once surrounded by a little band who have been there too, all eager to tell of the wonders of the country. The children's day begins at 6.30, and it is an hour before they are ready for breakfast, for, besides ordinary clothes, many of them wear splints and braces. At 7.30, the breakfast-bell rings, and all promptly answer to the call. The hours between breakfast and dinner are spent under the trees, — anywhere but in the house. Three go each morning and afternoon in the donkey-cart, and so a turn comes in time to each. Dinner comes at twelve, after which some of the weaker ones have a rest before going out. Tea is at five, then a bath and bed for the little ones. Those older remain up until eight o'clock. The children remain from two weeks to three months. While at the Home, all their clothing is supplied. At present twenty-eight children can be provided for. To do the work and take care of the children, there is one Sister in charge, two nurses from the hospital, a cook, a laundress, and a house-maid.

The house is built to accommodate many more children, but, owing to lack of sufficient funds, at present the Management can only care for twenty-eight. Their hope for the future is to be able to increase the nurses and servants, and so provide for more children.

The Association has supplied for The Children's Hospital, every year since its organization, all the bed-linen, the table-linen and towels, the clothing for the children, also boots, shoes, hats, and hoods, besides bandages and many surgical appliances. They give Thanksgiving and Christmas dinners to the household, a Christmas-tree to the children in the hospital, and through the year drives to such of the children as are well enough to enjoy them. The Association owns, and carries on entirely, the Convalescent Home at Wellesley Hills.

The number of children admitted to the Convalescent Home, —

1892 . . . 109 patients.
1894 . . . 162 patients from May to July 1.

E. A. W.

A LIST OF OFFICERS OF THE LADIES' AID ASSOCIATION AND CONVALESCENT HOME.

Presidents.

Mrs. ROBERT C. WINTHROP, 1869.   Mrs. O. W. PEABODY, 1892.

Vice-Presidents.

Mrs. W. W. TUCKER, 1869.   Mrs. O. W. PEABODY, 1879.
Mrs. E. L. DAVIS, 1892.

Treasurers.

Miss C. A. BREWER, 1869.   Mrs. H. HUNNEWELL, 1878.
Mrs. H. G. CURTIS, 1884.

Secretaries.

Miss M. L. ROBBINS.   Miss A. M. STORER, 1878.
Mrs. J. C. HOOKER, 1869.   Mrs. T. B. CURTIS, 1889.
Miss M. J. REVERE, 1874.   Mrs. H. C. WESTON, 1894.

OUT-DOOR RELIEF DEPARTMENT.

FOR some time there has been a need felt for a proper means of caring, at their own homes, for sick and crippled children who attend the out-patient department of the hospital. In March, 1894, an out-door relief department was organized, with a surgeon and a trained nurse in charge.

This department exists as an experiment, to determine, if possible, whether the nursing and care in their own homes, of sick and crippled children who are unable to come to the hospital, is feasible and desirable.

The patients which it is intended to help are: (1) Those children who are discharged from the hospital with dressings to be attended to, and who are too ill to be brought back to the out-patient department for this purpose. (2) Patients who come to the out-patient department and need that the apparatus which they are wearing should be regulated and looked after. (3) Those cases who are to be fitted to new apparatus after a period of bed-treatment at home.

In all these instances the nurse and surgeon visit the children in their homes, and the treatment is carried out which has been instituted. At the proper time, they are referred to the out-patient department. The parents and guardians of these sick children are often ignorant as to the application of apparatus. They are taught at home, and, with a little care, succeed in carrying out the directions much better than if instructed only once at the hospital, or out-patient department.

The treatment of the chronic patients which appear at the hospital must extend, necessarily, over many months or even years; and to make the parents of these children appreciate the importance of continued care, is an important part of the work of this department.

II. MEDICAL DIVISION.

THE Medical Division contains a series of original papers by members of the Staff. The paper by Dr. THOMAS F. SHERMAN was written shortly before his death.

TYPHOID FEVER.

BY F. GORDON MORRILL, M.D.

THE cases tabulated below are taken, in their con-
secutive order, from the records; the only qual-
ification necessary to be included in the series being
a moral certainty of the correctness of the diagnosis.
The extreme rarity of malarial fever in Boston renders
the diagnosis sufficiently easy in a large majority of
cases at this institution. From "simple continued
fever," typhoid is soon differentiated by a tender or
tympanitic abdomen, the appearance of rose spots or
undue prolongation of pyrexia.

It is not always so easy to distinguish between typhoid and the early stages of tubercular meningitis.
Nausea, or vomiting, fever, headache, anorexia, and constipation are symptoms quite common to the early stages
of both diseases.

A consideration of a fair number of cases of tubercular
meningitis which have been observed in the surgical
wards, and in which the first symptoms of the disease
were recorded, shows that night-cries, inequality of the
pupils, and the *tâche cerebrale* are signs on which the
most dependence can be placed in making a differential
diagnosis in doubtful cases. Moreover, an irregular or
intermittent pulse has been usually observed in tubercular meningitis on or before the sixth day from the
commencement of the disease.

TYPHOID FEVER.

No. of Case	Age	Sex	Normal morning temperature first noticed	Normal evening temperature first noticed	Highest temperature	Day on which highest temperature was noted	Quickest pulse	Rose spots first noted	Rose spots not noted	Nausea, or Vomiting	Bronchitis, or Broncho-pneumonia	Epistaxis	Diarrhœa	Acute delirium	Abdominal tenderness	Tympanites	Abdominal pain	Relapse	Day of relapse	Duration of relapse	Duration of Illness	Result	Remarks
1	10	M.	Day.		°				1		B.		1	1	1	1					?	D.	No history obtainable. Died of perforation a few days after admission.
2	4½	F.	18	20	105.2	10	140	10		N.	B.				1		1				41	R.	
3	10	F.			104	9	140		1	V.			1								102	R.	Passed bloody stools during nine days.
4	8	M.			104.3	6	120	8		V.	B.P.			1	1	1	1				75	R.	Vomited blood on day of admission.
5	4	F.			101	6	125	7			B.										39	R.	
6	7½	M.			104			14					1								25	R.	
7	7	F.	28	30	107.8	23	148		1		B.			1	1		1				62	R.	
8	10	F.			104.2	19	140	10			B.P.	1			1	1	1				77	R.	
9	8½	F.	23	29	103.8	6	112	9		V.				1	1		1				85	R.	
10	9	F.	21	31	101.8	9	140	5													83	R.	
11	7	F.	14	17	102	10	100	10		V.											37	R.	
12	9	M.	14	16	105	6		8													49	R.	
13	6½	F.	31	35	104.2		132	10		N.	B.P.						1				76	R.	
14	10	F.	38	38		33	128			N.	B.P.			1	1	1	1	1	41	21	115	R.	Tender and tympanitic abdomen during relapse. Highest temperature, 103° F. during relapse.

No.	Age	Sex			Temp.														Result	Remarks
15	5	M.	41	43	105	17		5	V.								B.		56	R.	During a period of one week the pulse intermitted every fourth beat.
16	6¼	F.	24	27	104.5	7	120		V.		1								67	R.	
17	4½	M.	28	31	104	10	120	21	V.	1		1				1	B.P.		57	R.	
18	3¾	M.	24	30	104	6		10	N.	1	1	1							47	R.	
19	11	F.	16	16	105	5	132			1	1								43	R.	Highest temperature during relapse, 102°.
20	7	F.	22	25	104.6	6	108			33	1						B.P.		70	R.	
21	12	F.	20	21	105.3	10	120	5		25	24	1	1		1		B.		91	R.	Highest temperature during relapse, 102.5°.
22	9	F.	24	26	104	9				21	29	1		1	1				48	R.	Highest temperature during relapse, 102°.
23	8½	F.	14	16	104	4	118	15				1	1	1		1	B.	V.	57	R.	Bloody dejections for a few days.
24	12	M.	28	54	104	10	112	14				1		1	1	1	B.			R.	
25	11	F.			103.6	6	128	16				1		1	1	1			50	R.	
26	8	F.	16	20	104.6	6	116	11				1		1	1	1			44	R.	
27	5	M.	13	15	104	5	136	6				1			1				42	R.	
28	9	F.	26	33	105	14		10			1	1	1	1	1				57	R.	
29	8	F.			104.8	22	160			8	39	1	1	1				V.		D.	Two relapses. 39–47. 50–55.
30	10	F.	10	10	104.6	6	128	6	N.&V.			1	1	1	1	1			27	R.	
31	12	M.	21	29	105.2	4	180		N.&V.			1	1		1				35	R.	Discharged on 9th day.
32	6	F.	25	26	104.8	7	128					1	1	1	1				46	R.	Probably tubercular complication.
33	12	F.			105.2	8	136					1		1	1				32	R.	
34	6	F.			104	9	128					1		1	1					R.	
35	12	F.	33	35	106	11	128		N.&V.		30	1	1	1	1	1	B.		35	R.	Discharged on 22d day.
36	6	M.	10	10	106.2	15	120		N.&V.			1				1	B.			R.	Discharged on 40th day.
37	9	M.	32		105	30	108		N.&V.			1							35	R.	
38	11	F.	22	26	105	16	116					1		1					36	R.	
39	6	M.	19	21	104.2	16	108	14				1		1	1	1	B.		40	R.	Under treatment at present date.
40	11	M.	33		104	12	114					1			1	1					
41	10	M.	33	29	104.8	21	198	21				1				1		B.		R.	
42	9	M.	23		105	10	194					1		1	1	1					Under treatment at present date.

Before making an analysis of the table below, a few words of explanation are necessary.

Normal temperatures are noted only when not due to the action of antipyretics. Under the heading " Bronchitis and Broncho-pneumonia " such cases are credited with bronchitis as coughed, but showed no signs of pulmonary consolidation. Very possibly, the cough in some instances may have been due to pharyngeal or laryngeal catarrh. Abdominal pain is noted only when voluntarily complained of during the first stage. Relapses are dated from periods of pyrexia lasting three days or more. This system is open to criticism, but is, I think, preferable to making relapses dependent upon the size of the spleen (no case being credited with a relapse unless the fever recurs after this organ has resumed its normal size), as being less liable to give rise to confusion. Enlargement of the spleen is not noted in the table, as the records in this respect are very imperfect. The " duration of illness " is considerably longer (on paper) than it would be in less liberally conducted institutions; for, in many instances, the parents are urged to allow their children to remain in the Hospital until their physical condition shall warrant their returning in safety to home diet and discomforts.

Average age, 8 +. Normal morning temperature first noted in thirty-one cases on the (average) 23d day. Normal evening temperature on the 27th day. Average highest temperature in forty cases (104.5° F.) observed on the (average) 11th day. Quickest average pulse (thirty-seven cases), 130 —, the range of all cases being from 100 to 180, the last figure that of a case dying from a tubercular complication. Rose spots first noted in twenty-four cases on the (average) 11th day. No rose

spots observed in 43% of all cases recorded. Nausea or
vomiting (or both) in 40%. Bronchitis (or rather cough
not due to pulmonary consolidation) in 36%. Broncho-
pneumonia in 14%, all of which recovered. Epistaxis in
29%. Diarrhœa in 33%. Acute delirium in 21%. Abdominal tenderness in 40%. Tympanites in 43%.
Abdominal pain complained of in early stage in 19%.
Relapses in 14%, of an average duration of seventeen
days. Time elapsing from appearance of first symptoms to date of discharge (34 cases), 55 days. Mortality,
a little under 5%. One case of house-infection has occurred, notwithstanding the care taken in disinfecting
the dejecta, and the scrupulous cleanliness which char-
acterizes the nursing under the supervision of the Sisters
of St. Margaret. One case deserves special mention :
here a temperature of 107.8° F., on the 19th day, fell in
a few hours to 101° ; this before the discovery of modern
antipyretics. Nothing beyond extreme prostration was
observed in this connection. Three days later, the
thermometer registered 104°, and then rapidly dropped
below normal, the patient showing a well-marked
tâche cerebrale and convergent strabismus, both of which
ominous symptoms disappeared in twenty-four hours,
and recovery slowly followed. The treatment at present
is very simple : milk-diet (or some form of predigested
starch), stimulants when indicated, and cool sponging
whenever the temperature rises above 102° F. Three-
hour charts not infrequently show a rise about noon,
and a fall before the evening temperature is taken. In
case cool sponging fails to reduce the temperature to a
comfortable range (an average reduction of one degree
is usually accomplished), phenacetin is given.

THE VALUE OF MILK LABORATORIES FOR HOSPITALS.

BY T. M. ROTCH, M.D.

THROUGH the generosity of certain philanthropists, I have been enabled, during the past summer, to make use of the products of the Walker Gordon Laboratory in the treatment of a great variety of diseases in both the out-patient department and the wards of the hospital.

It seems fitting that as The Children's Hospital was the first public institution in the world where this new and advanced method of not only feeding, but in this sense truly treating children and their diseases, has been carried out, it should have a place in the first medical report of the hospital. The idea of the milk laboratory is to provide milk in any form in which it may be ordered by physicians. That the milk as a whole should be freer from dirt, from bacteria, and thus from disease, than has ever before been accomplished in any city milk-supply; that it should be milk which has been received into and transported only in glass, and that from the time of the milking to the time of arrival at the laboratory only a few hours should be allowed to elapse. This milk, on its arrival at the laboratory, has been submitted to a careful bacteriological examination, and was found to contain bacteria in the thousands, rather than the usual millions in the c. c. commonly reported of the milk-supply of large cities. I have seen two

hundred gallons of this milk, filtered through a fine white
cloth, leave but a few black specks on the cloth. The
laboratory is fully equipped with the best apparatus for
manipulating this practically pure milk.

The three principal elements of this milk — fat, sugar,
and proteids — have, by a series of careful chemical
analyses, been so exactly determined that at any time
a physician's prescription can be put up with these elements, showing, within a fraction, any percentage which
he may call for. This latest triumph, in the province of
mechanical chemistry, opens at once to the physician a
vast field for investigation, and a great opportunity to
test, by exact methods, the different questions of infant
feeding, which in the past he could only judge of em-
pyrically, or, as I have had to study them, by incurring
the great expense of continual chemical analyses. The
laboratory, in fact, is presented to us as one more
instrument of precision to aid us in the scientific treatment of disease.

In the out-patient department, the physician in charge,
Dr. Wentworth, speaks of the inestimable good which
this milk, supplied on his prescription to the many young
infants who apply there for treatment, has accomplished
during the past summer; while in the wards I have tested
its efficacy with the remarkably good results which I
had already met with in private practice.

The money which for the summer months was gen-
erously given to supply this milk to the hospital, for
about a dozen children, has now been exhausted. It
is to be hoped, however, that the Managers of the hospital
will wisely consider the propriety of allowing the Staff
to prescribe for perhaps six or eight cases at a time,
when in their opinion such cases can be benefited by

this treatment.   The cases which, from my experience, require to be treated by milk modification are liable to be presented for treatment at all times of the year.   Perhaps the most simple way of illustrating the class of cases where this treatment is needed is to cite special cases, and to write the prescription for each.

Beginning in the out-patient department, where infants as well as children are presented for treatment, the most simple form of case for illustration is where a mother, Mrs. A., who is nursing her baby, and has plenty of good milk for it, as evidenced by her child's healthy condition, comes to see what food she can have her baby fed on, as she cannot afford to remain at home to nurse it.   The advice given is, of course, that her breast-milk is the best food for her baby, and that she must endeavor to continue to nurse it, if not at every feeding, at least three or four times in the twenty-four hours, and that it can have a substitute-food for the other meals. The analysis of average human breast-milk shows the following percentages : —

Fat 4.00
Milk-sugar. 7.00
Proteids . . . . 1 to 2.00

I have found, however, that although these are the average percentages, many healthy mothers are nursing healthy infants with the percentages of these three principal elements of the milk decidedly differing from the average percentages.   This is not only a fact, but I have also found that a healthy infant thriving on one of these more uncommon milks, may, when changed to an average breast-milk, be quite seriously affected, and continue to be so affected until it is again given the milk on which it was in the first place thriving.

Chemical analyses have, then, convinced me that not only the average breast-milk should be called a good milk, but also those milks which, although differing from the average, yet have been proved to best suit the especial infant. Bearing these facts in mind, we must endeavor to find, in those cases of infantile digestion which do not correspond to the average digestion, a breast-milk which, in its idiosyncrasy, corresponds to the especial infant's digestive idiosyncrasy. Now there is no means of knowing whether this mother above spoken of has an average breast-milk, except through a chemical examination of the milk; and yet, in preparing a food to supplement her milk, it is highly important to give her baby a preparation as nearly corresponding to the analysis of her milk as is possible. We may, of course, working somewhat in the dark, give a preparation corresponding to an average human milk; and this, before the laboratory was established, was practically what we had to do. We may, however, at once, by this empirical method, disorder a previously good digestion, and have to return to the breast again. The proper course of procedure is to make an analysis of the mother's milk before attempting to supplement it artificially, and then to closely copy her milk, whatever it is. In this especial case, the baby, a girl, was four months old, and it was estimated, by careful weighing before and after being nursed, that her average meal consisted of about four ounces. The analysis of the breast-milk showed that it differed considerably from the average, and that the baby would probably not have thriven on the percentages found in average breast-milks. The following figures will make the above remarks more easily understood : —

	I.	II.
	Analysis of Average Human Milk.	Analysis of Mrs. A.'s Breast-Milk.
Fat	4.00	3.00
Sugar	7.00	6.00
Proteids . .	1.50	3.00

The substitute-food which we should write a prescription for is then that shown in Analysis II., except that it is safer to begin with a lower proteid percentage, and the following is the form of prescription to be sent to the laboratory : —

℞

    Fat 3.00
    Milk-sugar 6.00
    Proteids 3.00

Send 4 tubes, each 4 ounces, once daily.
Lime-water, 1/20.
Heat to 167° F.
Direct the mother to nurse the baby 3 or 4 times in the 24 hours.

It will perhaps be well to explain here, that the addition 1 to 20 of lime-water makes the alkalinity of the laboratory milk correspond very closely to that of human milk. Also, that while a temperature of 167° F. is sufficient to kill any noxious developed bacteria which may be in the milk, it does not coagulate that fractional part of the proteids which is affected by a temperature over 171°, and thus the mixture, while being practically sterile, is yet not boiled. The possibility of having the milk heated to whatever temperature the physician may wish and order in his prescription, is one of the peculiar advantages of the laboratory's precise apparatus. Mrs. A. is now supplied with four meals for her baby, which she can safely have given it while she is away working for the money to pay for it. Many babies appear at the clinic, the victims of Infantile Atrophy from improper feeding. The treatment of these cases by cod-liver oil, both internally and by inunction, has always seemed to me to be as irrational as it was offensive. For years, in a pretty large experience with this class of cases during my service at the Infant Hospital, I have given no cod-liver oil. The disease is essentially a disturbance of absorption, and as often improves with a diminution of fat in the food as with an increase of that element. The treatment is to so lessen either one or all of the milk-elements that the absorbents will again perform their duty, and then to gradually increase the elements in their percentage until normal nutrition is established. In doing this, the fat percentage can, of course, be raised very high, if indicated, for the especial case. The laboratory is especially fitted to carry out the treatment of this class of cases. The following prescriptions are indicated in the various stages of the disease, from a much reduced condition, and almost entire inaction of the absorbents up to a condition of health and a perfect nutrition.

INFANT SIX MONTHS OLD, WEIGHT OF TWO MONTHS.

	℞ (1)	℞ (2)	℞ (3)	℞ (4)	℞ (5)	℞ (6)
Fat . .	1.00	1.50	2.00	3.00	4.00	5.00
Sugar .	4.00	5.00	6.00	7.00	7.00	7.00
Proteids .	1.00	1.00	1.50	1.50	2.00	3.00

The following prescriptions were written for cases which I treated in the wards.

CASE I. DUODENAL JAUNDICE.

Girl, six years old. Rapid recovery. No drugs given. Treated by relieving the duodenum from the digestion of fat.

℞
Fat	0.25
Sugar	6.00
Proteids	3.50

Send 6 tubes, each 6 ounces. 1/10 lime water, and heated to 167°.

CASE II. TUBERCULOSIS OF LUNG.

Girl. Age, 4 years. Vomits cod-liver oil when given. Very ill; apparently failing. Grew much brighter and stronger, stopped vomiting and had a return of appetite when cod-liver oil was omitted, and 8 ounces every four hours of the following modified milk-mixture given : —

℞
Fat	6.00
Sugar	7.00
Proteids	3.40

Heated to 167°.
Lime water; 1/10.

So far I have purposely avoided speaking of that large class of cases where an irritable stomach, in any form of disease, whether pulmonary or renal, medical or surgical, precludes the possibility of recovery, by not allowing sufficient nourishment to enter the body to keep the patient alive until the disease has run its course. In this class of cases, the careful regulation of the different percentages in the food, the administration of exact quantities of lime water, and the proper degree of heating, all of which can only be satisfactorily accomplished by laboratory manipulation, will, I am sure, in the future take the place of other methods of treatment, either by drugs or by food preparations either inaccurately compounded, or whose component parts are unknown and are hidden under a patent or proprietary name.

THE RELATION OF AN AURAL SERVICE TO THE NEEDS OF A GENERAL HOSPITAL FOR CHILDREN.

BY CLARENCE J. BLAKE, M.D.

A REVIEW of the aural statistics of the hospital, for the five years ending in 1893, gives a total record of forty-two cases, of which forty are included in the records of the out-patient department, and two in those of the house. These do not include, however, the cases in which the ear-trouble was secondary to, or a complication of, the disease for which the patient was admitted to the hospital, and in which, therefore, the description of the aural disease was included in the general record, and could be separated for statistical purposes only by an exhaustive reading of the records. Viewed from a practical point of provision for their needs, aural patients may be divided into the two usual classes of out and house patients, the former class, even among children, very largely outnumbering the latter, as is shown by the fact that, in one of the largest special hospitals for the purpose in this country, of all the cases of ear-disease there treated twenty-three per cent were children under fifteen years of age.

The maintenance, by any general hospital, of such special out-patient departments as those for eye and ear diseases must be determined by the demands for such service ; but in cities with a large and centrally located special hospital for this class of cases, it may well be questioned whether a general hospital is not unnecessarily complicating its general work by the maintenance of such special out-patient departments. In Boston, for instance, in regard to ambulant ear-patients, this would particularly apply, since we have a special hospital with provision for the accommodation of a large number of out-patients, a house service of twenty eight beds, and a ward for women and children, in which one hundred and eighteen children were cared for during the past year.

On the other hand, it should be said that the special out-patient department, with a continuous morning service, requires the daily attendance of the special surgeon, and his consequent availability for house-consultation and service at certain stated hours in each day. The importance of having an Aural Surgeon as one of the Consulting Staff of such an institution as the Children's Hospital has been sufficiently demonstrated during the last five years, and is evident on consideration of the frequency with which the ear is implicated in the course of the contagious diseases of childhood, the rapidity with which such implications run a destructive course in many cases, and the consequent importance of prompt attention.

In chronic general diseases of children also, the occurrence of an inflammation of the middle ear may complicate the case and call for expert examination and treatment; while in still other cases, an investigation of the ear, with negative result, may, by eliminating one possible factor, prove of service in the making of a correct diagnosis.

Much of the treatment in both the acute and chronic diseases of the ear in children may advisedly be left to an intelligent nurse, properly instructed, and the wards of the hospital afford an excellent opportunity for teaching. In the acute and chronic eczema of the auricle, for instance, attention to the application of astringent washes in the acute, and in the chronic of the various ointments, together with the care necessary in protecting the parts from external irritation, and the frequent dressings which are sometimes required, may properly fall within the province of the nurse, while the syringing of the ear, either for the removal of secretion from the external auditory canal, or for the cleansing which is an important part of the treatment of the suppurative diseases of the middle ear, should be considered a part of the nurses' regular course of instruction.

Furuncles in the external auditory canal in children are not of as common occurrence as in adults; but they are apt to be accompanied by swelling of the glands behind and below the ear and in the neck, and by considerable febrile disturbance, while the diffuse inflammation of the canal, which is of relatively more frequent occurrence is not only a source of much discomfort and of general disturbance, but sometimes leads to thickening of the skin and deeper tissues lining the canal, to a chronic eczema, and to the stoppage of the canal by keratotic masses, often extremely difficult of removal. The causes of this diffuse inflammation are various, from the mechanical irritation resulting from the introduction of foreign bodies or from injudicious attempts at their removal; from the presence of animal and vegetable parasites; from chilling of the ear, as, for instance, by the trickling of cold water into the canal during ice-water applications to the head, and, in the course of scarlet fever and measles, from the extension of the inflammation of the skin of the face to that of the auricle and canal. In measles also there is, in some cases, the occurrence of small serous blebs or hemorrhagic bullæ at the inner end of the canal and upon the upper border of the drum-head, the pricking of which affords relief from the pain incident to the vascular tension of this sensitive region. The great vascularity of the canal, the drum-head, and the middle ear in children favors the rapid course of an inflammatory process, and, in a child suffering from scarlet fever or measles, with other sufficient evident cause for rise of temperature and complaint of pain, a discharge from the ear is often, in default of the aural inspection which should be considered imperatively as a part of the examination of all such cases, the first evidence of an implication of the ear.

A serous oozing from the drum-head and external canal sometimes simulates, to superficial observation, a suppurative discharge, the serum mixing with the small particles of exfoliated epidermis and the ceruminous secretion, and having the appearance of thick yellowish pus.

The external auditory canal in the young child is principally cartilaginous, the bony canal, which in the adult forms the inner half, being formed partly by development of the osseous tympanic ring, and partly by the projection of the mastoid cells outward, forming the posterior wall. The facts that deficiency in bony development is not uncommon, and that the vascularity and delicacy of the soft tissues favor inflammatory invasion of the surrounding parts, emphasizes the importance of keeping a strict watch upon all cases of disease of the external and middle ear in children, a not uncommon complication to be borne in mind being the simple post aural abscess, which, if not speedily treated surgically, results sometimes in extensive denudation of bone, superficial caries, and still more serious consequences.

The most frequent diseases of the ear in childhood are those occurring in the middle ear, either secondarily as the result of extension of inflammation from the external ear, or from the naso-pharynx through the Eustachian tube, the latter by far the more common channel for the invasion of the middle ear, and the primary cause being usually either the simple congestion and swelling of the mucous membrane of the nose and naso-pharynx, such as may occur in the course of an ordinary head cold, the presence of the so-called adenoid growths in the naso-pharynx, or the inflammation of the mucous membrane of this cavity, the Eustachian tube, and middle ear, incident to the course of the exanthemata.

In the first and second instance, the middle ear is affected primarily as the result of the closure of the Eustachian tube, either by the swelling of its lining-membrane or by the pressure of the adenoid growth, the results being interference with the ventilation of the ear and with its blood-supply, and a congestion of the middle ear, which in coryza is usually more acute than in the cases where the pressure of the adenoid growth produces a static congestion. In the "earache of childhood," which is most commonly the result of a simple acute inflammation of the middle ear, attention must be directed, not only to the organ specially affected, but to the nose and naso-pharynx and to the general condition of the child, — a faulty diet or faulty habits of life being often found to be the predisposing causes; while, in the cases of adenoid growths, early surgical treatment, thoroughly carried out, is the best radical measure. One of several objections to the partial and frequently repeated operation without ether for the removal of adenoids is, that, while the central growth may be thus disposed of, the region of the Eustachian tube, which needs especially careful manipulation, is left comparatively untouched, and, while there may be satisfactory evidence of the establishment of nasal breathing, the passage of air through the Eustachian tubes to the middle ears is hindered by the presence of the remaining lateral portion of the adenoid growth; therefore, with reference to relief of the aural symptoms, the complete operation, under ether, permitting thorough and careful manipulation of the tubal region, is preferable.

The differences between the structure of the middle ear in children and in adults should be borne in mind in the consideration of the inflammations occurring in this cavity; for, while in the adult the walls of the middle ear are formed of compact bone or firm membrane, with comparatively little vascular communication with other parts, in the child the vascular anastomoses are large and numerous. The divisions of the temporal bone do not assume their relationship of complete ossification until after puberty; the mastoid process, which subsequently forms the posterior wall of the osseous canal, and presents in its interior a mass of diploetic and pneumatic cells, the latter communicating with the middle-ear cavity, is represented in the young child by the mastoid antrum, in direct communication with the middle ear, and immediately beneath the thin plate of bone, later known as the tegmen mastoideum, which forms the boundary between this and the cranial cavity, while the sutura petroso-squamosa permits direct communication, through membranous connections, largely supplied with blood vessels, between the lining-membrane of the tympanum and the meninges.

The late Dr. Edward H. Clark said, very justly, " So necessary is a careful attention to the ear, during the course of an acute exanthema, that every physician who treats such a case without careful attention to the organ of hearing, must be denominated an unscrupulous practitioner."

The importance of frequent examination of the ears during the course, especially in the acute stage, of scarlet-fever and measles, is emphasized by the fact already stated, that the middle-ear disease, occurring as a complication of these affections, usually runs a rapidly destructive course, and by the further fact that aural symptoms, other than objective ones, unless unusually severe, are apt to be lost sight of in the consideration of the general condition of the child.

A careful review of the causes of deaf mutism, including in the enumeration those cases in which the cause was doubtful, show twenty-seven per cent to have been the result of suppurative inflammation of the middle ear in scarlet fever.

Another form of middle-ear disease demanding watchfulness is that which occurs in connection with tuberculosis in children; because the disease, beginning usually in the upper portion of the tympanic cavity, runs its course almost painlessly, the first evidences objectively being a slight congestion of the inner end of the canal and upper border of the drum-head, followed speedily by œdema and rapid destruction of tissue, with at first a thin and watery and then a copious purulent discharge; the whole course of the primary stage of this

affection, from the first congestion to the establishment
of a copious purulent outflow, being in some cases
included within the space of twenty-four hours.

The statement which has been made as to the opportunities afforded for the training of nurses in the care of
cases of ear disease in the hospital wards, holds true in
an equal or greater degree to the medical student and
house-officer; indeed, the latter, if he has not had the
necessary preliminary training, should be instructed, at
the beginning of his service, in the methods of examining
the ear, and in the diagnosis of those conditions which it
is most important for him to recognize, since it is the
house-officer who has the patients under continuous
observation. To further this end, it is proposed to
offer, in the Aural Department of the Massachusetts
Charitable Eye and Ear Infirmary, a short special
course on examination and diagnosis for the house-
officers of The Children's Hospital, as a preliminary
to their entering upon their hospital duties.

MALARIA IN CHILDREN.[1]

BY EDWARD M. BUCKINGHAM, M.D.

M ALARIA is a new disease in Boston. Although it has existed in Massachusetts at three, probably four, different times since the settlement of the country; yet, until within a few years, it has been unknown since about 1836, and at that time it scarcely reached the eastern part of the State. Furthermore, it is increasing. For both these reasons, the cases in this hospital seem worthy of more study than would otherwise be due to so small a number. It is my object to ascertain the forms in which it appears in a district so recently and so slightly affected by it; also, how it differs, if at all, when seen in children, from the same disease as found among adults; and I start on this inquiry without any preconceived opinions.

The hospital has the records of but twenty-three cases, not including one of doubtful diagnosis, but which was, I think, probably malaria. These cases were found both in the wards and in the out-patient room. Seventeen, *i. e.*, over seventy per cent of the whole, were received in the past twenty-one months; while of all the patients with all diseases in the medical services admitted in ten years, but twenty-six per cent were

---

[1] This paper was prepared in 1892, in anticipation of an early publication of this report. The writer, however, sees no occasion for making any change, except to state that, for the present, the epidemic seems to have passed its height in this city.

received in the same time. This tends to show that the disease is on the increase with us. The same thing seems to be shown in my own service at the Boston City Hospital, in which sixteen cases, mostly of home origin, were admitted in the summer just past, against nine in the corresponding months of the previous year. The earliest case was admitted ten years ago,[1] malaria being known to exist in the city somewhat earlier.

Twelve or thirteen patients lived in Boston; five or six in Brookline, from which place the hospital receives many patients, most of these last coming from near the Boston border. The remainder were from suburban places; two had previously had malaria in other States. The exact locality of some of the suburban cases is not stated; but all the other patients came from low damp ground or from higher land on its borders, and all the Boston cases came from districts reported to be malarial in the State Board of Health Report for 1889.

The ages of the patients whose cases are included in this Report are shown in the following table : —

Ages.	No of Cases.	Ages.	No. of Cases.
2 years	1	8 years	3
3½ "	2	9 "	2
4 "	1	10 "	4
4½ "	1	12 "	1
5 "	1	12½ "	1
5½ "	1	13½ "	1
6 "	1	? "	1
7 "	2		

It should be understood that patients are only received in the wards between the ages of two and twelve,

[1] No cases were admitted in January or February, and but one in March, seven in April, May, and June. eight in July, August, and September, six in October, two in November, and one was received in December, but dated from the middle of November.

inclusive; while in the out-patient room there is no minimum age, and the upward limit is somewhat elastic.

Contrary to common experience, more girls than boys, fifteen to nine, had malaria. If this is not merely an accident due to a small number of cases, it tends to show that, in childhood, males are not subject to greater exposure, as is the case in adult life.

In two or three cases tonsilitis had preceded or accompanied the first malarial symptoms. Occasionally the onset was very sudden and alarming; but one patient had been ill for a week before any paroxysm, and often the approach must have been gradual. In no other way can I account for the existence of symptoms for a month, five weeks, eight weeks, and ten weeks before applying for treatment, either at the hospital or elsewhere. In some of these cases, merely malaise had been observed by the parents; but often there had been symptoms which should have been recognized by a competent observer.

Chill existed in some degree, at some time, in all the cases except three, or perhaps four. In one of these, the diagnosis was made by tertian fever without any first stage, but followed by sweating, the type changing to quotidian after the fourth paroxysm, great pallor, and enlarged spleen. Another patient, without chill, had a daily fever followed by sweating, the temperature falling to normal in the intervals, the spleen very large. In both cases, the use of quinine was followed by rapid recovery.

In another case, the diagnosis depended solely upon the temperature-chart and the effect of treatment. Since this diagnosis may be questioned, the case is reported. The child, six years old and apparently well, was ad-

mitted to the hospital for prolapse of the rectum of long
standing. Previous to operation, the thermometer was
used as a matter of routine and registered 104.5°. Every
night the temperature rose to from 104° to 105°, and
every morning it fell to normal. This lasted for eight
days, until quinine was given; and after this there was no
more fever. There were neither abdominal tenderness,
nor rose spots, nor diarrhœa. On the third day, because
of a suspicion of redness in the tonsils, the child was
transferred to the medical service and isolated; but this
redness was extremely slight and temporary. There
was no reddening of the mucous membrane elsewhere,
nor were there any other symptoms of scarlet fever. The
girl was kept under observation a long time for desquamation; but it did not appear. I have examined
a large number of scarlet-fever charts without finding
one resembling this. The child came from a locality
known to be malarial. Enlarged spleen was not
detected. A fourth case will be spoken of later.

All the others had chill; yet it differed greatly in
degree and constancy, being sometimes nothing more
than coldness, or coldness with pallor, while in four
cases it amounted to rigor, not necessarily in every
paroxysm, but in one or more. That the difference in
degree of different chills in the same individual did not
depend on treatment alone, was made evident through
the unwillingness of the staff to treat malaria before the
diagnosis was quite sure. It was perhaps held in reserve, at times, longer than would have been the case in
a more thoroughly malarial region. In about two-thirds
of all the cases, chills are said to have been slight, or
are merely mentioned as chills. From the best information I can get from recollection, from consulting other

members of the staff, house-officers, and Sisters in charge
of the wards, it seems certain that in the majority, though
not in all the cases, the chill was never severe, but
amounted merely to sensations of coldness. The duration
of chills varied from a very short time up to two hours.
So far as these records show, the age of the child has some
influence on the chill; those being known to be severe,
occurring, one in a child of three and a half years, the
others at ten or older; thus, as the age advances, the
type of chill has a tendency to approach that in
the adult. Chilliness of some sort existed in the child of
two years. As this was the youngest in the series, there
is no opportunity to verify the statement of Forcheimer,
that this feature of the paroxysm is the one most often
omitted in infants under two. Mildness of the chill, in
this series, is due to the age of the patients, and not to
mildness of the epidemic; for the adults observed by me
in the City Hospital had well-marked chills, as a rule.
The common want of a well-marked chill, as in adults,
or its replacement by other nervous symptoms in the
child, is not peculiar to malaria. Again, some patients
had chills, such as they were, with every paroxysm;
others omitted certain chills, although enough phenomena were present to make it sure that the paroxysm
existed. Thus, one patient had a convulsion with his
first paroxysm, this being replaced by coldness in those
following. Another patient, who often vomited with the
chill, occasionally vomited without it at the regular
hour, the case being otherwise well marked. Yet another child sometimes omitted the chill, but had in its
place headache and epigastric pain. Although chill was
sometimes slight throughout, yet it by no means followed
that this was an index to the severity of the case, or even

of the paroxysm. One child of four years, who had chilliness but no rigor, yet ground his teeth and had constant micturition, frequent dejections, and vomiting while cold. This was followed by fever of several hours. The child's appetite was quite good on alternate days, yet in two weeks it had lost weight and color. The spleen was enlarged.

Convulsions occurred in one patient of four years, already mentioned. Here the diagnosis was made plain by daily chills for two weeks, accompanied by pain. They occurred also in a child of nine, who had very irregular chills for a month before admission. Both patients had large spleens, and responded quickly to treatment. In neither case was there more than one convulsion; in the first, it lasted for half an hour, the child being white and cold at the time. Convulsions occurred also in a case about the diagnosis of which there may be some reasonable doubt, and which is reported somewhat fully in order to show the possible difficulty of making a diagnosis between this disease and tubercular meningitis. The decidedly tertian type is to be noticed. The child, four and a half years old, lived near other cases of undoubted malaria, in a wooden building over an unpaved driveway. He had had an abscess of the jaw from injury, and had several enlarged glands, had some evidence of rachitis, especially in the head, had been losing weight and color, had disturbed sleep and poor appetite, was said to be irritable every second day, but had neither chill nor sweating. The spleen was not enlarged. Temperature on the evening of admission rose from 100° to 103.°8, with vomiting during the rise, and fell to normal in the morning. Two days later it rose to 104.°8, with vomiting, thirst, restlessness, and partial delirium, and again fell in the morning, the child becoming quite comfortable. He was then given quinine, gr. vi., during the twenty-four hours. Again, in two days from the last paroxysm, quinine being continued, the temperature rose to 102°, with headache and vomiting, and again fell to normal, the child being quite weak after this paroxysm. Quinine was continued, with an additional eight grains before the attack. On the expected day, the temperature rose, but earlier in the day, and only reached 101°, falling to 97° afterward. Quinine was then omitted. Three days later, the temperature rose to 103°, with headache, followed by delirium, nystagmus, strabismus, and convulsion, dropping to 99° in the next morning. Again, after two days, were vomiting with nystagmus, but without fever, and then another day of comparatively good condition; to be followed by another, with headache, restlessness, and vomiting, the temperature reaching 101°. Then followed some time of poor condition, with a slightly irregular temperature, to be interrupted by scarlet fever. During the course of this last disease, the temperature never exceeded 102°, but, during the time that the rash was fading, there were two nights of delirium, for which no cause was assigned, and these nights were separated by one of good sleep; the course again following the tertian type. There were at no time signs in the chest, nor enlarged spleen. Two months later, he had fever, with consolidation in one lung, both of which quickly passed off. He then moved to a suburb not as yet malarial, and became strong and well, remaining so for a year and a half. The regular tertian course of the sickness and the effect of quinine were both marked. That he was going to recover could not be foreseen, and would not absolutely exclude tubercular meningitis, if it could.

Delirium occurred, but only in the case just reported was it unaccompanied by convulsion in the same paroxysm.

Vomiting occurred with more than a third of the patients, and one other complained of nausea. Vomiting occurred in one or in more paroxysms, either with or without a chill, but always as part of a paroxysm, if at all. Its severity bore no relation to that of the chill.

Incontinence of urine and of feces occurred during the paroxysms of one child; and incontinence of urine alone, perhaps occurred in another, although, in this last case, the relation of cause and effect could not be absolutely traced.

Headache was almost universal in children of nine or over, but was not recorded at all in those younger than that, except in a boy of four and a half. Of course this difference was very likely due to the inability of small children to make known their sensations. One child described its headache as frontal; the others did not locate it. It occurred either in the chill or after it. Sometimes, in individual paroxysms, it replaced chill. Other pains were situated in the neck, the arms, legs, liver, spleen, or epigastrium, the latter decidedly predominating. Sometimes tenderness occurred without other pain, and once there was œdema at the same time, — a combination which might easily be taken for rheumatism.

Fever was nearly universal. It is recorded in all but three cases, all of whom were out-patients in whom the thermometer could not be used. One only of the three asserted definitely that there was no fever; but this child had both chills and sweating, and fever may have been overlooked. There were several temperatures of from 103° to 106°, and occasionally slight irregularities for a few days, to be followed by a sudden rise; but, owing to great rise and fall, the thermometer is deceptive unless it is used often. Merely two readings in the day may fall far short of giving a true impression. Duration of high temperature was once noted as lasting several hours.

Sweating was recorded in about a third of the cases. It may have been present in some others, but certainly not in all. It was reported at three and a half years, but in general was more often found in the older children.

Many children were wearied, and some exhausted, by the paroxysm; several went to sleep as soon as it was over; but there were none of the interesting cases described by Holt,[1] to which attention was first called by uncontrollable fits of periodic sleepiness in school-children, and in which the diagnosis was confirmed by the thermometer, no other symptoms of the paroxysm existing. On the other hand, insomnia throughout the disease occurred in another case.

In nearly half the cases, the paroxysm occurred daily, five were tertian and one quartan, while several were more or less irregular; thus, a paroxysm every second or third day, without regularity, or a quotidian case, with a paroxysm occasionally omitted, or a change of type from quartan to tertian. Quotidian cases were more numerous than all others previous to 1891. Since

[1] Symptoms and Diagnosis of Malaria in Children. L. Emmett Holt, M.D., New York, 1883.

then there has been an increase in the proportion of
tertian cases, but this is probably accidental; for, in the
near suburbs, while Dr. E. H. Stevens, of Cambridge,
has met mostly with quotidian in the proportion of two
to one of tertian, Dr. G. K. Sabine, of Brookline, has met
only with tertian cases.

The hour of the paroxysm varied in a quarter of the
cases from two P.M. to three, five, six, or later; while,
in three quarters, it was in the morning, or was not
recorded, — showing that too much reliance must not be
placed on the time of greatest fever in differentiating
this disease from typhoid. The hour also is subject to
change, as well as the day; one case having a chill in
the middle of the night at first, but afterward in the
morning, — a fact, the knowledge of which may be
important, in case treatment does not give the expected
results.

The spleen was found to be enlarged in fourteen
cases, a little less than two-thirds; it was sometimes
made out by percussion, sometimes by palpation, and
sometimes in both ways. The largest spleen extended
from the seventh rib to an inch and a half below the
ribs; and there is every sort of record short of that. It
was once described as hard. Children do not always
like to have their spleens examined, and may be counted
on to object unless treated with gentleness. Even when
they do not object, it is not always easy to find it;
especially as, often, they cannot be persuaded to assist
the examiner by deep inspirations. Therefore the fact
that enlargement of the spleen cannot be found, does not
prove its absence. Its presence, on the other hand, of
itself limits the diagnosis to probably malaria or typhoid,
although not certainly so. Determination of the spleen

by palpation is probably more reliable than by percussion. Various writers[1] state that the child's spleen is
normally larger in proportion than that of the adult.
Nevertheless, as we cannot detect it by palpation in
the state of health, it may be taken as enlarged when
we do feel it.

The liver was observed to be enlarged but twice;
once it was tender, as well.

Disturbances of digestion, with coated tongue, anorexia, foul breath, constipation, or diarrhœa, were noted
in many cases. These symptoms were somewhat permanent; and yet one tertian case was noted, in which
appetite was good on alternate days; and, in another,
frequent dejections, as already stated, formed part of the
paroxysm. Most, but not all, of the patients were anæmic or debilitated after a short time, some sallow or jaun-
diced; but none of these symptoms existed in every
case. In general, cases of long duration were much
exhausted; but exhaustion was not invariably propor-
tioned to the duration of the disease. Other symptoms,
observed in individual cases or a few cases at most,
were herpes, epistaxis, and slight cough, with a few
râles in the chest.

From the above analysis, it appears that malaria has
a somewhat different type in children from that which it
presents in adults, and that, in general, the older the
child, the more likely is his malaria to approach the
adult type. The paroxysm of children is more likely to
be wanting in completeness, one or another of its stages
being omitted. While many cases had two stages, less

---

[1] Largest at 5 years. Canteteau, quoted by Holt. Ibid. p. 13. Largest
towards the age of eight. Gaston and Vallée. Revue Mensuelle des Mala-
dies de l'Enfance, Sept. 1892, p. 398.

than a quarter are recorded as presenting a complete paroxysm of three stages, most of these being among older children. Even in this small proportion, one or another stage was often slight, although others might be well marked. The various paroxyms in a given case are not always alike, even when untreated; and there is a decided tendency for the cold stage to be very mild, or replaced by other nervous symptoms, such as vomiting, delirium, or convulsions. Coldness, however, can often be detected by careful examination. Headache is not to be made out in the most of the young children, although it very likely exists. Other pains may exist as well. Fever is the most constant single symptom, and the thermometer ranges from a slight rise to at least as high as 106°; but with us, as yet, it falls to normal in the intermission. Paroxysms may be severe, or may be so mild as to attract little attention. These peculiarities are for the most part due to the age of the patient, and not to the mildness of the epidemic; for they are not found in anything like the same proportion among adults in the same city. The spleen will probably be found enlarged; but the non-detection of this symptom does not prove its absence. Enlarged liver has been observed, and, in a small proportion of cases, cough. The diagnosis from typhoid presents especial difficulty, both diseases having a type in children different from that seen in adults, and somewhat resembling each other. The temperature-chart must include frequent observations. While the time of the rise is generally in the morning, yet, as has been shown, an evening rise by no means excludes malaria. A fall to absolute normal, in the intermission, is, however, suggestive. It may not be out of place to quote the warning of Forcheimer,[1] that, where doubt exists, obvious prudence requires precautions to be taken, as in cases of certain typhoid. The following case is reported in some detail, because, while the tertian course and general history leave no doubt as to diagnosis, yet at first it might have been taken for the beginning of convalescence from acute pneumonia or from typhoid. A boy of twelve was admitted, July 14th, with the story that he had been sick for a week, but had not been in bed. He had a very poor appetite, very foul breath, and a severe headache. The bowels had been moved twice in a week by Epsom salts; at entrance he had diarrhœa, and was quite weak and weary. There were slight cough and profuse sweating. There were no rose spots. He was well nourished, but somewhat pale. The temperature was 106°, and had been higher. The pulse was 128. On July 15th, a few questionable fine râles were detected in the centre of the left back. The chest was otherwise negative. The spleen was felt for one and a half inches below the edge of the ribs, and its percussion dulness extended to the seventh rib. Urine was clear, acid, with s. g. 1014, no albumen, and Diazo reaction absent. The temperature fell to normal this morning, and the boy seemed better. There were a few small spots of herpes on the upper lip.

July 16th, there was a chill, and at noon the temperature was 104°. The edge of the liver could be felt on full inspiration. Headache, and pain in the back of the neck existed with general malaise. There were no rose spots. The tongue was dry, and covered with a brown coat. The lips were markedly cyanotic after the chill. On the 17th, epistaxis occurred. On the 18th and 20th,

[1] Keating's Cyclopædia, Vol. I., p. 893.

there were chills. He was then given quinine, and made a rapid recovery.

In this series, there were two or four relapses, as one chooses to consider them. One patient had had malaria three years, and another one year before. A third was said to have a relapse from malaria in New York, and the fourth relapsed while at the Convalescent Home.

Treatment was with arsenic in one chronic case. It was given in the form of Fowler's solution, in doses of six drops throughout the day, to a girl of nine. She recovered in three weeks. All the other cases were treated with quinine, sometimes in divided doses during the day; sometimes in one, or practically one, large dose six hours before the chill; and sometimes the two methods were combined. I think that no one, even if unfamiliar with the literature of malaria, could study these records without forming the opinion, that to give quinine in divided doses spread over the day is to make a comparatively inefficient use of it in the treatment of this disease. The following table shows the doses of quinine given in some of the cases that have most quickly recovered : —

Age.	Dose.	
3½ years,	. . 3 grains.	in one dose.
6 "	. . 7 "	" " "
8 "	. . 4 "	in four doses, with an hour's interval.
8 "	. . 7 "	
9 "	. . 15 "	in three doses.
12 "	. . 12 "	in three two-grain and one six-grain dose.

In no case is there any record of quinine being vomited, or of its causing any unpleasant symptoms, unless, possibly, headache once or twice, — which headache may have been due to the disease. In no case was quinine preceded by a cathartic. These doses may seem large to some physicians of this vicinity. They are small as compared with doses given in some places, but they were sufficient. It is not unlikely, however, that we may have to increase our doses, if the vicinity becomes more malarial.

Administration has been commonly by pill, which thus seems to be efficient when only moderate doses are needed. Quinine may be given in coffee to older children; but it is not a great improvement in taste, and there are objections to giving coffee to children. It should never be given in milk; for, although the taste is somewhat disguised if the milk is fresh, yet there is danger of disgusting the child with his food. Chocolate tablets of tannate of quinine have not been given, but they are stated by Forchcimer to be efficient in malaria; and they are not disagreeable, which is an important point in treating children. The following way of giving quinine, provided the dose is not large, is a good one. It is not original here. Cut a cone from a confectioner's chocolate cream. Rub up with some of the cream a half-grain of quinine and pack it into the opening, which is then to be closed with the remainder of the cone. This is rather agreeable than otherwise to many children, provided always that it is eaten as soon as prepared, but if kept, it soon becomes bitter.

To sum up. The clinical picture of malaria differs somewhat in children, especially in children under ten, from that seen in adults. The peculiarities are a less clear marking of the different stages of the paroxysm, and especially of the first stage, in which there is a tendency for the chill to be slight, or to be replaced by other nervous phenomena. Fever is the most constant symptom.

In this series, it has been necessary, in making the diagnosis, to consider the possibility of the acute exanthemata, tubercular meningitis, lobar pneumonia, and typhoid, — possibly, also, lobular pneumonia and rheumatism. Lastly, treatment in this freshly invaded community, has not required so large doses of quinine as are sometimes used in regions where the disease is better established.

AN EPIDEMIC OF SCARLET FEVER

BY THOMAS F. SHERMAN, M. D.

DURING the winter of 1891–92 scarlet fever was very prevalent in Boston. In the six months from October 1, 1891, to April 1, 1892, 1492 cases were reported to the Board of Health, as compared with 594 cases for the same period of the preceding winter.

The type of the disease has varied, some of the cases being very mild, while others were more severe.

The average mortality rate in Boston from scarlet fever during the last five years has been 7.21 per cent. During the six months from October to April the mortality rate was 6.03, — showing that the disease as a whole was rather less severe than the average.

The epidemic which forms the subject of the present paper occurred in two children's homes in this city over which I have the medical care. These homes contained, during the period of the epidemic, 180 children, coming from the lower grades of society, placed in the homes either temporarily or for adoption, and ranging in age from six months to fifteen years, — the majority of them, however, being from four to ten years of age. Of these 180 children, 30 of them, or 16.6 per cent, were attacked with scarlet fever during the progress of the epidemic. One child had the disease twice, making 31 cases in all. All of the children were treated by myself during their illness, with the exception of five, who were kindly received at the City Hospital; but whose histories, for the sake of completeness, I have included with the others, through the courtesy of their attending physicians.

ORIGIN OF THE EPIDEMIC.

We are rarely able to accurately trace the source of an epidemic. There is always the possibility that the infectious germ may have entered through some undiscovered channel; yet, in the present instance, the origin of the disease can, I think, be fixed with a very fair degree of probability, as follows : —

Agnes D., aged 5, entered Home, which we will call No. 1, on Monday morning, November 2. She came from the scarlet fever ward of one of our hospitals in the city, where she had been sick with the disease since August 13. Desquamation had ceased three days previous to her discharge from the hospital. She had been bathed and thoroughly disinfected before leaving. She had no sore throat or chronic otorrhœa. Clean clothes throughout had been carried to her from the Home. She was received into the Home Monday morning, November 2, and on Friday morning, November 6, two children broke out with the scarlatinal eruption, and two more children were taken sick on November 7. The last previous case of scarlet fever in the Home had occurred August 12.

It seems fair to presume that Agnes D. in some way brought with her the germs of scarlet fever into the Home. It may have been that the desquamation, which had apparently ceased, began again. This has occurred in several of my later cases.

Whether this second desquamation is contagious or not I do not know. I believe, however, that a child should be kept isolated as long as any desquamation remains, and that all convalescents from contagious diseases who have been treated in hospitals should be boarded outside for a period of two weeks before being received into an institution and allowed to mingle with other children.

The history of the outbreak of scarlet fever in Home No. 2 was as follows: —

On Wednesday morning, February 24, three children went from the Home to a Dental College for the purpose of having their teeth examined. While there, they met in the waiting-room a number of other children. They did not leave the Home again. On Monday night, February 29, two of the children complained of headache with sore throats, and on Tuesday morning they were found well covered with the eruption of scarlet fever. The third child showed no symptoms till Thursday night, March 3, when she complained of sore throat and vomiting, with the appearance of the rash on the following morning, March 4. No previous case of scarlet fever had occurred in the Home since June 10, 1891. On the supposition that the children contracted the disease from contact with some child at the Dental College, we have a period of incubation of $5\frac{1}{2}$ days in the cases of two of the children, and of $8\frac{1}{2}$ days for the third child, if she contracted the disease at the same time with the other children, or of $2\frac{1}{2}$ to 3 days if, as is more probable, she took it from them.

AGE. — Of the 31 cases of scarlet fever occurring in the two Homes, the ages were as follows: —

Under 2 years . . . . 0 case.   From 5 to 10 years . 19 cases.
From 2 to 3 years . . 1 "   From 10 to 15 " . . 7 "
From 3 to 5 " . . 4 cases.

Sixteen of the cases were females and fifteen were males.

In both of the Homes, by adopting strict precautions and by prompt isolation, the disease was prevented from gaining an entrance into the nurseries occupied by the youngest children. This had never been possible in previous epidemics of measles or whooping-cough.

DURATION OF STAGE OF INCUBATION.

In considering the duration of the stage of incubation, only those cases have been selected where the time of infection could be very positively fixed. For example, on the morning of November 13 a child was received into the Home from a small village where there had been no cases of scarlet fever. Five days later, on the morning of November 18, the child had a convulsion, followed by the scarlatinal eruption two hours later.

Again, on January 4 and 5, two children were received coming from an institution free from the fever, and in each case the prodromal symptoms appeared after an interval of three days.

In ten cases the period of incubation was fixed, and the duration was as follows:—

DURATION OF STAGE OF INCUBATION.

3 days in	2 cases.
3½ " in	1 case.
4 " in	1 "
4½ " in	1 "
5 " in	3 cases.
5½ " in	2 "

Average period of incubation was 4.4 days.

Symptoms and Duration of Prodromal Stage.

In many of the cases the prodromal symptoms were so very light as to escape notice, or were wanting altogether, the eruption being the first thing to attract the attention of the nurse when the child was given a bath or was being undressed for bed.

Duration of Prodromal Stage.

Absent or unnoticed in	12 cases.
Less than 5 hours in	2 "
5 to 10 hours in	4 "
10 to 15 " in	8 "
15 to 20 " in	2 "
20 to 25 " in	2 "
41 hours in	1 case.

Average duration in 19 cases, when it was noticed, 13.4 hours.

In the case where the prodromal stage lasted 41 hours, the child had an attack of vomiting one evening at 6 o'clock; temperature, 103°. The next day she seemed very comfortable, with moderate fever, and it was not till the day following, at 11 A. M., that the eruption of scarlet fever appeared.

Prodromal Symptoms.

Vomiting occurred in	16 cases.
" absent in	15 "
Sore throat noticed in	20 "
" " absent in	11 "
Both sore throat and vomiting in	11 "
Neither sore throat nor vomiting in . . .	6 "
Epistaxis in	1 case.
Persistent diarrhœa in	1 "
Convulsions in	1 "

Duration of Fever.

By this is meant the length of time from the appearance of the rash, before the temperature became normal, in simple uncomplicated cases, or in those cases where the complications occurred during convalescence.

3 days in . . . . 1 case.
4 " in . .   . . 1 "
5 " in . . . . 2 cases.
6 " in . . . . 3 "
7 " in . . . . 1 case.
8 " in . . . . 4 cases.
9 " in .   . . . 1 case.
10 " in . . . . 5 cases.
12 " in . . .   . 4 "
15 " in . . . . 2 "

The average duration of fever in 24 cases was 8.7 days.

Desquamation, — Commencement and Duration.

Desquamation began as a rule when the eruption had faded and the temperature became normal. The earliest date noticed was on the fourth day, the latest on the 37th day. It was usually lamellar in character, though in some cases it was furfuraceous throughout, as in measles. Its extent seemed to depend upon the severity of the eruption. The duration of desquamation varied within wide limits. In five cases it apparently ceased for over a week, and then began again on the fingers and feet. It would be important to know if this second desquamation carries with it the dangers of infection to other children.

DESQUAMATION COMMENCEMENT.

On the 4th day in . . . . 1 case.
On the 5th " in . . . . 1 "
On the 6th " in . . . . 3 cases.
On the 7th " in . . . . 4 "
On the 8th " in . . . . 7 "
On the 9th " in . . . . 2 "
On the 10th " in . . . . 4 "
On the 11th " in . . . . 3 "
On the 16tb " in . . . . 3 "
On the 24th " in . . . . 1 case.
On the 37th " in . . . . 1 "

Seventy per cent of the cases began to desquamate from the 6th to the 12th day.

Leaving out the two cases where desquamation was abnormally delayed, the average date of commencing desquamation was 8.9 days in 28 cases, which closely corresponds to the average duration of the fever in 24 cases, which was 8.7 days. The very slight desquamation which often occurs around the corners of the mouth forty-eight hours after the appearance of the eruption was not considered in the above table.

DURATION OF DESQUAMATION.

For 17 days . . . . 1 case.
For 21 " . . . . 1 "
For 26 " . . . . 2 cases.
For 30 to 35 days . . . . 3 "
For 35 to 40 " . . . . 6 "
For 40 to 45 " . . . . 4 "
For 45 to 50 " . . . . 2 "
For 50 to 55 " . . . . 5 "
For 61 days . . . . 1 case.
For 92 " . . . . 1 "

In 26 cases where the desquamation was carefully noticed, the average duration was 41.6 days.

COMPLICATIONS.

Cervical adenitis	16.1 per cent.
Rheumatism	16.0 " "
Nephritis	12.8 " "
Otitis media	12.8 " "
Stomatitis ulcerosa	6.4 " "
Diphtheria . . . .̇	6.4 " "
Varicella	1 case.
Chorea	1 "

ADENITIS. — This occurred in five cases. No cases are included under this head where the glandular swelling was the result of inflammation of the fauces, diphtheria, stomatitis, or otitis. In two cases fluctuation was detected, and the swelling lanced. In three cases absorption took place without further trouble. In two cases it was bilateral, and in three cases single. Intermittent high temperature accompanied the swelling in all cases. The time of its occurrence was as follows : —

On the 5th day in .	. . 1 case.
On the 7th " in .	. 1 "
On the 18th " in .	. . . 1 "
On the 20th " in .	. . . . 1 "
On the 25th " in	. . 1 "

RHEUMATISM, FIVE CASES. — This was only considered to be present in those cases where we had marked swelling and tenderness of the joints, with elevation of temperature, or where after a sudden rise of temperature an endocardial murmur was found which had not been present previously. In three of these cases, the joints alone were affected, wrists and ankles, without the heart being complicated.

In two cases, after a sudden rise of temperature to 103°, systolic murmurs were found at the apex transmitted into the back. In these cases the joints were not affected.

In one case the murmur had disappeared after an interval of three months. The other case was unfortunately lost track of.

One case of endocarditis was followed by a severe case of chorea five days later. The child had, however, a previous history of chorea before having the scarlet fever.

The three cases of joint affections occurred on the 8th, 11th, and 12th days of the disease.

The two cases of endocarditis occurred on the 20th and 27th days.

NEPHRITIS, FOUR CASES. —

TIME OF OCCURRENCE.

On the 12th day in . . . . . . 1 case.
On the 14th " in . . . . . . 1 case with a relapse
                              on 24th day.
On the 23d day in . . . 1 case.
On the 24th day in . . . . . 1 case.

In one case the nephritis occurred two days after the child had been given a warm bath, and might have been caused by the skin being unduly exposed. The other cases occurred in children who had been kept in bed and confined to a bread-and-milk diet. In the majority of cases, nephritis, like the other scarlatinal complications, is probably of microbic origin. A rise in temperature of one or two days usually preceded any change in the urine. In all cases the urine was diminished in amount, and contained a large amount of albumen, with blood and casts. In no cases was œdema noticed to any extent. In one case persistent vomiting was a prominent feature. All made good recoveries.

OTITIS MEDIA, THREE CASES. — This occurred with otorrhœa on the 18th, 19th, and 23d day. Paracentesis was done in one case. In two cases the otorrhœa ceased in three weeks. In one case it is still persistent to a slight degree, three months after it started.

STOMATITIS ULCEROSA, THREE CASES. — These occurred on the 19th, 20th, and 21st day of the disease. One case only was severe, accompanied by marked fetor and extensive sloughing of the mucous membrane of the cheek, glandular swelling, and albuminuria.

DIPHTHERIA, TWO CASES. — True diphtheria occurred as a complication in two children, aged 4 and 6 years, on the 19th and 20th days. They were convalescent from the scarlet fever, with normal temperatures, for over a week.

It was not possible to account for the origin of the disease, unless it sprang from the unavoidably crowded condition of the sick room at that time. The disease was severe, spreading over the pharynx and invading the larynx, causing a total loss of voice in both cases.

The youngest child, aged 4, died from exhaustion on the 18th day. The other child recovered, after an illness of three weeks. In this case, paralysis of the soft palate, with regurgitation of liquids, followed.

VARICELLA. — This occurred in one case, after the child had been isolated for fourteen days, and was desquamating, showing a period of incubation for the varicella of not less than two weeks.

ALBUMINURIA. — Albuminuria was present during some period of the disease in 13 cases, or 41.9 per cent. Dr.

Dudley, in his observation of 100 cases of scarlet fever at the City Hospital ("Boston Medical and Surgical Journal," February 11, 1892), found it in 49 per cent. Leaving out the cases where it was caused by the different complications, or by a resulting nephritis, it seemed to be dependent upon the elevation of the temperature.

MORTALITY. — The mortality rate for the 31 cases was 3.22 per cent; the single death being caused by diphtheria, as noted above, on the 38th day. The average rate in the city of Boston, during the last four years, has been 7.21 per cent, and for the six months from October to April 1, covering the period of the cases above recorded, was 6.03 per cent. The average rate in 1000 cases reported by Dr. Caiger ("London Lancet," January 6, 1891) was 4.67 per cent, and of the London Hospitals, during the last five years, is 9.63 per cent.

SECOND ATTACK, ONE CASE. — This case is recorded not alone on account of its being a second attack, but because the second attack followed so soon after the first one.

FIRST ATTACK. — Arthur A., aged $5\frac{1}{2}$ years, entered Home No. 1 on Friday, November 13. He was well till Wednesday morning at eight o'clock, when he had a convulsion, followed two hours later by a well-marked scarlatinal eruption; temperature, 104.°6; pulse, 160. No vomiting or sore throat. He was transferred to the City Hospital, and came under the care of Dr. G. B. Shattuck, to whose kindness I am indebted for the rest of his history. His fever was of moderate severity, and lasted for ten days. Desquamation began on the seventh day, and continued for 48 days. Otitis media, on the right side, occurred on the 19th day, and he had an attack of endocarditis, with mitral systolic murmur, on the 27th day of the disease.

On January 12, after being in the hospital nearly seven weeks, he returned to the Home.

SECOND ATTACK. — He remained well till the night of February 1, 19 days after his discharge from the hospital, when he complained of headache and vomiting. The temperature rose to 103°, and on February 2, at 4 P. M., the rash of scarlet fever appeared for a second time. This attack was also of moderate severity. Desquamation began on the 16th day, and continued for 61 days. His right ear was again affected, otorrhœa occurring on the 23d day. He escaped the attack of endocarditis, however, and careful examination of the heart showed that the murmur found in his first attack had disappeared.

THE ETIOLOGY OF CHOREA.

A STUDY OF ALL THE CASES OF THIS DISEASE IN THE OUT-
PATIENT DEPARTMENT FROM 1883 TO 1891.

BY CHARLES W. TOWNSEND, M. D.

IN the following paper I have endeavored to analyze the records of all the cases of chorea in the out-patient department of the hospital, with reference particularly to the etiology of the disease and its relations with rheumatism and endocarditis, — all important points in any consideration of its prevention and treatment.

In a shifting out-patient clinic it is often difficult to learn the ultimate results of treatment and the course of the disease, especially in one of such prolonged duration as chorea. The records of the out-patient service are, however, somewhat unusual in this respect, as many of the patients come back, year after year, and, by means of indices to the record books, their first record is found, and subsequent data added. The record books themselves, unlike the books often found in out-patient departments, which give but one line for name, diagnosis, and treatment, devote two folio pages to each case, the important headings being printed on one of these pages.

Where, however, the patient has been lost sight of, I have endeavored to hunt him up, and thus learn the present condition of health as to chorea, rheumatism, and endocarditis. For this work I am greatly indebted to Dr. F. A. Higgins, late house-officer, who performed this task with much skill and industry. In order to allow a sufficient time to elapse, no case has been included later than December, 1890.

The total number of cases of chorea was 148, out of a total of 4,080 cases of all kinds, or a percentage of 3.6. Of these 61 were male, 87 female; a preponderance of females in the proportion of almost 3 to 2. As the patients were all under the age of puberty, it seems probable that the preponderance of females is not due to any sexual difference in itself, but to different constitutions or habits of life. Thus, among boys there is greater robustness, both inherited and also acquired by their more vigorous out-of-door life. The following table gives the ages, and it will be seen that 112, or 76 per cent, were between the ages of 8 and 12 years, the youngest being 4 years, the oldest 15 years old.

Age.	No. of Cases.	Age.	No. of Cases.
4 years	2	10 years	20
5 "	4	11 "	25
6 "	6	12 "	26
7 "	10	13 "	6
8 "	23	14 "	5
9 "	18	15 "	1

The disease is said to be rare among the colored, and it is to be noted that only one of these patients was of negro blood, although the colored are frequently seen in the clinic. Six cases were recorded as unilateral.[1]

Fifty-nine cases, or about 40 per cent, had more than one attack, eighteen having had three attacks, and six as many as four attacks, while two have had six attacks, coming on in one case every spring.

---

[1] All cases of post-hemiplegic chorea or athetosis are of course excluded from the list.

A study of the months in which the cases occurred, as given in the following chart, is of interest, as showing the relation of this disease to school-work.

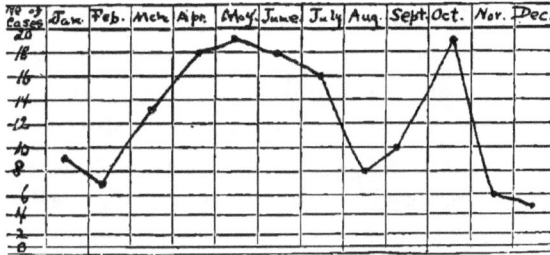

This chart is not very unlike that constructed by Dr. Lewis, of Philadelphia; where the greatest number of attacks occur in March, instead of May; and where, also, there is no rise in October. His chart corresponds with an acute rheumatism chart, and with a storm-centre chart for the city. My chart would be intelligible on the assumption that the mental pressure at school, with its accompanying physical depression, act as exciting causes for the disease.[1] As a consequence, the results begin to show in the last half of the school year, in March, by the greater number of cases of chorea. In April, May, and June, when the strain is greatest, and the children are undergoing examinations to determine promotion, the cases of chorea abound, and in July, after the school is over, a nearly equal number apply for treatment. During August and September, they fall off in numbers, while in October, a few weeks after school has begun again, there is a fresh burst of cases, many of

[1] In this way Sachs, in his article on chorea, in Keating's Cyclopædia of Children's Diseases, interprets the monthly variations of this disease. He too finds an outbreak of attacks on the opening of the schools in the autumn.

these being due to the lighting up of previous attacks. These children being withdrawn from the schools, the disease is again comparatively infrequent till March.

Many of the teachers in the public schools send a child away as soon as the St. Vitus' dance is discovered, telling the parents to consult a physician. It is unfortunate that this rule,—which should be as strict as in cases of the exanthemata, but for a different reason, — is not universally followed, for the child is often allowed to remain until the disease becomes so marked as to oblige removal from school. The reason of this delay is due, first of all, to an inability to appreciate, in the cases of gradual onset, the significance of early slight symptoms, often attributed to nervousness or to a habit of fidgeting. Early symptoms may appear to the ill-informed teacher as evidences simply of a restless and untidy disposition, and the child is scolded for its poor handwriting, caused by a disease beyond the control of its will. In some cases the delay in discharging the child from school is due to the fact that the teacher as well as the parent is loath to have a child that has been pushed forward, and stands high in the class, lose the chance of promotion by leaving school for what seems to them an unimportant ailment.

This is a story which is frequently told, and its importance as an exciting cause of chorea cannot be doubted, and serves to emphasize the evils of over-pressure in schools, and of the struggle for rank and promotions. If teachers and parents were instructed on this point, by timely removal of the child from school, they would prevent the occurrence of many cases of chorea.

One case is recorded as having been cured of an attack

of chorea by removal from a public to a private school, where the pressure was less.

Fright is generally stated to be in some cases an exciting cause for this disease. But five cases are attributed to this in the record books.

Another exciting or reflex cause for the disease, closely connected with school work, has been found in eye-strain, due to some imperfection in the eyes, especially hypermetropia and astigmatism. In eleven of the cases, — chiefly among the recent cases, as in the earlier ones, this point was not considered, — some defect was found. Eight of these were corrected by glasses. For most of this work, the hospital is indebted to Dr. F. E. Cheney, and some of these cases will be found in detail in a paper read by him December 11, 1889.[1]

Of these eight cases, five were cured or very much benefited, while three were not improved.

These few cases hint, it seems to me, at a very important matter in connection with chorea, namely, that chorea, in predisposed cases, especially if it has occurred before, may be started up by some reflex irritation, as by the strain on imperfect eyes, or, as in one of my cases, by the reflex irritation from an earache, or as Dr. Jacobi has shown, by nasal irritation. A tight prepuce, a bound down clitoris, or lumbricoid worms are other examples of reflex causes. Secondary attacks, brought on by reflex causes, may be explained in the same way that secondary attacks of whooping-cough, brought on by a subsequent bronchitis, are explained; namely, that the local irritation causes in a reflex manner a discharge of nervous action along the same well worn channels as in the

[1] " Errors of Refraction and Insufficiencies of the Ocular Muscles as Causes of Chorea, with Cases." Bost. Med. & Surg. Journ., Feb. 20, 1890.

original attack. Hence secondary attacks may not always be true chorea as they are not true whooping-cough.

Looked at in this way, it would be absurd to suppose, as some have felt inclined, that all cases of chorea could be cured by correcting the ocular defect, or to go to the other extreme and deny any merit to this procedure.

Excessive tea-drinking, so often seen among this class of patients, and giving rise to a well recognized group of symptoms, appeared in one case to be partly responsible for the chorea.

A debilitated state of the system was present in nearly all the cases; while there were a few exceptions, where the patient was apparently robust. The number of cases seen in an out-patient clinic, on the other hand, which are free from chorea, and yet debilitated, and often by lack of out-door exercise and over pressure at school, show us that something else than these causes are at work in the cases of chorea.

A nervous constitution, due to anæmia or to neurotic inheritance, was not uncommon, and certainly seems to predispose to chorea.

In two cases the trouble appeared in girls at the time of the establishment of the menstruation, which was irregular and difficult.

In five cases, a brother or sister of the patient suffered from chorea at the same time, with or at a greater or less distance in time from the case in question. This might be explained by imitation, the brother or sister assuming a nervous trick. This theory is untenable, except in rare instances; for, if this were so, the same trouble would occur among the children in the open wards of the hospital containing cases of chorea, and this is not the case. A more reasonable supposition is, that, being of the same family, the children have similar tendencies, and are probably exposed to the same exciting causes.

We now come to the interesting and important consideration of the relation of chorea to rheumatism and endocarditis, and although the later hospital records are generally complete, it is to be regretted that many of the earlier records are deficient in this respect, so that the frequency of these two affections is undoubtedly understated. Many of the earlier records were, as was stated above, completed by the recent review by Dr. F. A. Higgins. The following tables show the results of the study of the relations of rheumatism and endocarditis to chorea in these cases.

TABLE I.

	No. of Cases	Per cent of total number.
Chorea	148	
Heart murmur	44	30
Rheumatism	31	21
Organic heart disease	21	14

TABLE II.

	Total.	Rheumatic.	Non-rheumatic.
Organic heart disease . .	21	15	6
Functional or transitory .	10	1	9
Undetermined	13	4	9
Total cases	44	20	24

Among an equal number of non-choreic children of the same ages, I found that eight had had rheumatism, and ten "growing-pains." This gives us a percentage of $5\frac{1}{2}$ rheumatic, or, making a liberal inclusion of "growing-pains," a percentage of 12. As will be seen by reference to Table I., 21 per cent, as a low estimate, were found to be rheumatic among these cases of chorea, This, however, leaves over three-quarters, or — supposing a great overlooking of rheumatism — certainly one-half the cases of chorea free from rheumatic history.

The intimate connection between chorea and rheumatism seems to be well illustrated by one family. The mother is subject to rheumatism; out of eleven children, three have had chorea, two of these having had rheumatism, one with endocarditis and permanent heart injury in addition. The third choreic patient had heart disease; but no history could be obtained of articular rheumatism. Of the remaining eight non-choreic children, one only has had rheumatism, and she also has heart disease.

Rheumatism in some of the cases seemed to be most closely connected with the attack of chorea; the rheumatism ceasing, to be immediately followed by chorea. In other cases rheumatism followed the chorea. Thus, in one of the cases in the list, which afterwards died of heart disease, the patient first had chorea; six months later there was found to be organic disease of the heart, and several months later an attack of rheumatism occurred, accompanied by purpura. Here it seems as if the chorea were responsible for the endocarditis; but it is to be remembered that rheumatism in children is often overlooked, and cases which result in permanent valvular damage may have had so little joint pain that the parents disregarded the trouble entirely, or classed the slight pains under the misleading head of "growing pains." It seems to me that it is far safer, from a therapeutic point of view, to consider all so-called growing pains as cases of rheumatism, and to remember that in children endocarditis and permanent heart damage may accompany very slight articular pain. The readiness with which many cases of "growing pains" yield to treatment with salicylate of soda is significant.

Another case is worth relating here as an example of heart disease following chorea, without any suspicion of rheumatism at any time. Lilly Q. was first seen at the clinic in July, 1883, at the age of nine years, suffering from her second attack of chorea. The heart was then normal. The first attack occurred two years before, immediately after she had been frightened by a dog. The third and last attack occurred when she was eleven years old. Within a year after this she became much debilitated and short of breath, and presenting herself at the clinic in June, 1888, she was found to have a mitral systolic lesion. She was from that time under frequent observation; the heart gradually enlarged; the mitral lesion became obstructive as well as regurgitant, and when last seen, in January, 1890, her condition was one of advanced heart disease. At no time in her career did she have rheumatic pains or anything that could be so interpreted, neither had she suffered from scarlet fever or diphtheria.

It is interesting to note that there were five other cases like this, where chorea was followed by organic disease of the heart without the occurrence of articular rheumatism; and there were a number of other cases of organic disease following chorea where rheumatism occurred for the first time after the heart was damaged.

As will be seen by the tables, thirty per cent of the cases had a heart murmur; twelve of these cases proving organic, three, to my knowledge, having since died. Of the remainder, more than one-half were undetermined, not a sufficient time having elapsed under observation to

prove whether they were organic or not. My experience
in these cases has, however, led me to look with suspicion on so-called functional heart murmurs.

Where the patient has had rheumatism, the origin of
the endocarditis is plain. Where no rheumatism has
occurred, how can we explain the endocarditis?

Some say that endocarditis may, in children, be the
sole manifestation of rheumatism, hence the heart lesion
of chorea is always rheumatic. Although this may be
true in some cases, it does not seem to me that we are
justified in making so sweeping a statement. Others
assuming, and with considerable justification, that chorea
is a specific infectious disease, would say that endocarditis followed it as it does other infectious diseases, like
scarlet fever and diphtheria. However this may be, the
frequent occurrence of endocarditis is certainly a strong
argument against a simple neurotic theory.

The transient cases of murmur can, it seems to me, be
explained in two ways, both of which may be right.
First, that the transient murmur is of anæmic origin, or
due to a regurgitation, through a mitral orifice relaxed
by the debilitated condition of the patient; or, secondly,
that a mild endocarditis exists, from which recovery is
practically if not pathologically complete.

What is this connection between chorea and rheumatism?

Is it simply a coincidence owing to the fact that
chorea, a nervous symptom, is apt to occur in debility
from rheumatism, as in debility from any other cause?
Or is there an intimate relationship between these two
possibly infectious diseases?

It seems to me that the latter view is the correct one,
because : —

1st. The proportion of rheumatic cases is too large for mere coincidence.

2d. Chorea is often followed or accompanied by rheumatism, as well as preceded by it, one sometimes giving place to the other.

The frequent association of chorea and endocarditis cannot be used as an argument in favor of the rheumatic view, for it is only begging the question and assuming that the endocarditis is always a manifestation of rheumatism.

Whatever the exact relations of chorea and rheumatism may be, it seems to me that a study of cases like these teaches us the very practical lesson that in a choreic child we should be wide awake to any indefinite pains as evidence of rheumatism, and should treat them accordingly; and that we should be particularly watchful for endocarditis. Furthermore, we should, as Osler has said, be as careful of a child with a murmur following chorea, as we should if that murmur followed acute rheumatism.

These cases also teach us that although the reflex and exciting causes enumerated above may start up chorea in non-rheumatic children, we should be particularly alert to avoid or remove these causes in the rheumatic, and above all in those who have had previous attacks of chorea.

As a further summary of this study, I would draw the following conclusions : —

1st. Fright, eye-strain, debility, and school pressure, particularly the latter, which often includes some of the former, are potent exciting causes of chorea.

2d. Rheumatism, although absent in at least half of

the cases of chorea, occurs with greater frequency among the choreic than the non-choreic cases.

3d. There is an intimate relationship between chorea and rheumatism.

4th. The heart murmur so frequently found in chorea, with or without the history of rheumatism, is in a considerable proportion of the cases due to endocarditis, and leads to organic valvular disease.

AN ENDEMIC OF DIPHTHERIA APPARENTLY STOPPED BY THE USE OF ANTITOXIN.

BY F. GORDON MORRILL, M. D.

DURING 1894 there were three outbreaks of diphtheria at the hospital, which rendered the institution worse than useless so far as the wards were concerned. Applications for admission had to be refused. Those who had the disease were transferred to the contagious wards, and all the other children were hurried away to the Convalescent House, at Wellesley, or to their own homes. The wards in which the cases had occurred, together with the furniture and bedding, underwent the usual process of cleaning and disinfection, involving much trouble and expense, aside from the grave inconvenience of having the working of the hospital thrown out of gear for days at a time. The statistics of the disease were very unsatisfactory, as they have been everywhere previous to the use of antitoxin. It has also been very unsatisfactory, to say the least, to have a child enter the house for the purpose of having bow-legs straightened or an eczema cured, and then die of a diphtheria contracted in the wards. It was thought at first that the disease was brought in by the parents and friends of the patients, and blouses and other precautionary measures were made obligatory to such visitors, in the hope of rendering them (externally) sterile. But in spite of these measures, the thing recurred, and an active investigation (which is now in progress) will result, I think, in remedying certain faults

in the ventilation, and thus reduce future danger to a
minimum. The disease has apparently originated, every
time, in one particular ward. But be this as it may, the
result has always been the same; for all the children
who were well enough to walk, or even to be carried, to
the top of the building, met in the playroom every day,
and medical and surgical patients have suffered alike.

The cases were usually ill-defined at the start; a child,
suffering from what was apparently nothing more serious
than a cold in the head or a mild tonsilitis, would fatally
infect others, or perhaps suddenly develop well-marked
diphtheria in his or her own throat. In December, there
were two mild cases, both of which I injected with antitoxin, and would have then used the remedy throughout
the wards, as a prophylactic, if it had been obtainable;
but everybody at that time was hard pressed to obtain a
sufficient quantity for a single case in private practice.

January 13th, eight days since, there were nine cases
in the contagious wards, three of which were serious.
During the preceding forty-eight hours, there had been
four or five cases sent over from the house, and the out-
look was anything but cheerful. It was not my term of
service; but as the staff very kindly gave me *carte blanche*
in the emergency, I took charge, so far as the diphtheria
was concerned, and used antitoxin very freely, and with
particular reference to its immunizing properties. Every
patient in the house was given an injection of five cubic
centimetres of the serum prepared by the Pasteur Insti-
tute of New York, and admissions were continued with
the proviso that each child should be given a like amount
upon entrance. The patients in the contagious wards
have been sent back to the house, after being disinfected,
whenever they have shown two negative cultures three

days apart. No fresh cases have occurred; and if we may judge by the history of previous outbreaks in the hospital, the present condition of affairs would appear to be the very satisfactory effect of a perfectly defined cause, — it being the opinion of the staff that, if protection had not been afforded, the hospital would have been temporarily closed before now, and for the fourth time within twelve months. The patients who received the dose for immunity were suffering from various ailments, medical and surgical. No exception was made for any cause. In all, 50 were injected, including those in the contagious wards. Three nurses and 39 children were injected for immunity. The ill effects were unimportant: well-marked urticaria, on the fifth or sixth day, in four; ephemeral diarrhœa, within twenty-four hours, in seven; temporary increase of micturition in nine; anorexia and slight vomiting in four; well-marked tenderness over site of injection, and extending up to axilla, in two; erythema around the puncture in three. All these minor evils disappeared promptly. In some cases, an increase of appetite, and immediate improvement in general condition, were noticed by Mr. Herbert J. Hall, the surgical interne, to whom I am indebted for good work in connection with the subject of these very hasty notes. These were mostly suppurative cases, — spine and hip. Some very sick children were injected. In the surgical wards, the serum was given in the evening (in many cases as late as 11 o'clock).

In no case to which I shall refer was the temperature normal when the injection was given. The surgical charts show an average temperature of 99.8° F. Four hours later, it was 99.4°, the slight drop being accounted for by the hour (between 12 and 3 A. M.) at which it was

taken, when the minimum would naturally be reached.
Of those who showed a reaction (21 cases), the thermometer registered an average rise of .7° ten or twelve hours
after the injections, and this in the morning, when, in
children sick from other causes, the temperature would
naturally be lower than in the evening. The greatest
rise in any case was 1.8°. In one there was a drop of
2.8° ; in one of 1.2° ; in another of 4°. In a single instance, there was no change whatever. These four were
spine or hip patients. In 19 of the cases which reacted,
the evening temperature averaged 99.6°, — a remission
of about .5° at a time when, in a sick child, a rise would
usually be expected.

The temperature of 13 medical cases injected for immunity was 99.4°. These were all treated in the afternoon. Four hours later the average was 99.5°. Of the
seven who reacted, the morning temperature (say sixteen
hours after treatment) showed a rise of a small fraction
less than a degree, — this at a time when it would naturally be less than in the afternoon. Two cases registered
a drop of .4°, one of 1.4°, and another of 1.2°. In one
case, no change was noted. A lobar pneumonia, which
had just dropped to normal, showed the greatest reaction
so far as temperature was concerned, a rise of 2.4° ; but
it dropped back to 99° in the evening, and there was
absolutely no disturbance of its general condition. The
average evening temperature of the cases which reacted
was 99° in the evening. In short, the reactions were
unmistakable, and agree with what has been hitherto
observed.

Of those treated for diphtheria in the contagious wards,
there is not much to be said. Two of the sickest children
were sitting up in bed and enjoying themselves twenty-

four hours after receiving a dose of 15 or 20 c.c. In both cases, the temperature dropped twenty-four hours after treatment, — one from 104° to 101.2°, and the other from 100.4° to 100°. Both were "nasty" cases. The worst case was sent over with a croupous pneumonia and the worst kind of diphtheria. In addition, there was almost a suppression of urine, and edema of face and legs. In all, 45 c.c. were injected in this instance; but the child died sixty hours after the first dose, the temperature showing little or no change until a few hours before death, when it dropped from 102° to 101°. At the autopsy, a broncho-pneumonia (from infection by the staphylo-coccus) in the lung opposite the original trouble (croupous pneumonia), was found. It was very extensive, and would have killed in any case. It is the only instance of an unmistakably double infection of the kind that I have ever seen or heard of. The mild cases have all done perfectly well, and three have been disinfected and returned to the general wards. There were no urticarias noted in the contagious wards.

JANUARY 23, 1895.

III. SURGICAL DIVISION.

THE Surgical Division consists of three parts : —

PART I.— A description of the work done in the surgical wards of the hospital, in the surgical out-patient department, and in the surgical appliance shop.

PART II. — A description of the standard routine of procedure in the classes of cases treated at the hospital. In selecting those to be emphasized, the editor has ventured to adopt a principle that he believes to be of value to the true progress of medicine, this is, the reporting in detail the treatment of the most common diseases, rather than recording unique or rare affections. The articles are signed by the initials of the writers.

PART III. — Original surgical papers on subjects selected by members of the Staff.

PART I.

THE RELATIVE FREQUENCY OF SURGICAL DISEASES AT THE HOSPITAL.

THE following is a list which represents the relative frequency of the cases : —

CLASSIFIED LIST OF MEDICAL AND SURGICAL DISEASES,

1869-1893.

Anæmia	92	Meningitis, Tubercular	44
Asthma	32	Nephritis	116
Bronchitis	220	Paralysis, Cerebral	72
Cholera Infantum	78	" Infantile	55
Chorea	210	Pericarditis	19
Debility	92	Peritonitis, Tubercular	55
Dysentery	94	Pertussis	51
Eczema	218	Pleurisy	115
Eneuresis	88	Pneumonia	133
Epilepsy	52	Psoriasis	24
Fever, Scarlet	52	Rhacitis	156
" Typhoid	88	Rheumatism	91
Gastro-intestinal Catarrh	308	Rubeola	82
Heart, Valvular disease	198	Stomatitis	71
Hydrocephalus	21	Syphilis, Congenital	28
Malaria	35	Tuberculosis, Pulmonary	113
Marasmus	27	Unclassified	215
Meningitis, Cerebro-spinal	18	Varicella	63

Abscess	264	Dislocations, Traumatic	29
Adenitis	178	" Hip, Congenital	39
Adenoids	62	Empyæma	52
Angioma	30	Fracture	281
Anomalous development	20	Genu Valgum	218
Atresia Ani	15	" Varum	427
Cellulitis	18	Lateral Curvature Spine	142
Concussions	52	Necrosis of bone	63
Conjunctivitis	28	Paralysis, Cerebral	185

8

Paralysis, Spinal 300	Tuberculosis, Ankle 300
Paronychia 20	" Elbow 15
Periostitis 20	" Hip 1402
Phimosis 91	" Knee 104
Prolapsus Recti 51	" Meninges . . . 20
Pyæmia 15	" Peritoneum . . 15
Rhachitis 345	" Phalanges . . . 28
Synovitis 88	" Shoulder . . . 15
Sarcoma 15	" Skin 15
Sprains . . 124	" Unclassified . . 84
Syphilis 25	" Vertebræ . . . 1964
Talipes Calcaneus . . 15	" Wrist . . . 20
" Equinus 29	Vaginitis 26
" Equino Varus . . . 397	Wens 20
" Valgus . . . 127	Wounds 30
" Varus . 28	Web Fingers 20
Torticollis . . . 79	

DESCRIPTION OF WORK IN THE SURGICAL WARDS OF THE HOSPITAL; IN THE OUT-PATIENT DEPARTMENT; AND IN THE SURGICAL APPLIANCE SHOP.

THE surgical work of the hospital is divided between two services. Each of these is subdivided into an in-door and out-door department. The services are classed as first and second surgical services, or with the subdivision of first in-service and first out-service, second in-service and second out-service. Each service is under the charge of the same surgeon throughout the year, who has superintendence of both the in- and out-patient departments of his service, although the actual charge of the out-patient department is delegated to an assistant surgeon, who also serves throughout the year. The in- and out-door departments are so associated that they form one and the same service so that the practice recommended in one is, as far as practicable, carried out or continued in the other. This is for the purpose of securing a uniform system of treatment, — a great advantage in the management of chronic cases lasting for a year or more; so that, under this plan, a child with hip or spinal disease is practically under the same supervision throughout the entire period of treatment. During the course of the disease, the patient may several times enter the hospital, and be discharged to the out-patient department, without any change in the system of treatment.

The two services, although conducted independently, are in close touch with one another, and no change in the routine of the work of the hospital is made except after consultation between the surgeons. Once a month, a tour of inspection is made by the surgeons in each other's services, that harmony of practice may exist.

IN-SERVICE.

The surgeons make daily visits to the hospital; direct all treatment, which is to be carried out under the direction of the house surgeon; order all apparatus, and attend to the detail of the application of appliances in cases demanding such treatment. Appliances are ordered only by the surgeon in charge.

At the surgical visit, all orders are given to the house surgeon, who is instructed to write them legibly in an order book provided for that purpose. No change in the treatment of a case is made without the approval of the surgeon, or the house surgeon in his absence; but at any time the house surgeon may be suspended by the Lady Superintendent, and the surgeon sent for, at the Superintendent's discretion.

The house surgeon measures for all apparatus, and, having made out the specifications, hands them to the Lady Superintendent, who arranges with the parents of the patient for the payment of the apparatus. It is then ordered. It is sent back, from time to time, from the surgical appliance shop and tried on, under the supervision of the surgeon, until it is fitted properly, and completed. Then the child is allowed to be about in the ward and playroom of the hospital for a short time, when it is sent home, instructions being given to

GIRLS' SURGICAL WARD.

the parents by the house surgeon as to the application and management of the apparatus, and when the patient shall come to the out-patient department; and a discharge blank (see page 132) is sent filled out to the assistant surgeon in charge of the out-patient department.

A case is discharged from the hospital only after its name has been entered in the discharge book, with the signature of the visiting surgeon. In case a patient is discharged otherwise, the person discharging it signs the discharge book and assumes the responsibility of the patient leaving the hospital. As soon as it is possible, the patients are removed from bed and placed upon a rolling table, or allowed to sit up in a rolling chair. As soon as practicable after the morning visit, the children who are able are sent to the playroom, where they are amused during the day under competent nurse supervision. Any case in which a suspicion arises of a contagious disease is immediately isolated, by the advice of the house surgeon and direction of the Lady Superintendent, in a room provided for that purpose. After the diagnosis has been confirmed by the surgeon under whose charge it is, it is transferred to the isolated wards, where it becomes a medical patient, with an assistant surgeon detailed to advise with the physician in charge of the case as to its surgical treatment.

Once a week, the surgeon inspects the surgical records of his service, and visits the out-patient department, and sees such cases in consultation as may have been saved for him by the assistant surgeon.

Each service in the house has assigned to it two operating days a week, on which the operations of that service are conducted, priority being given to it on those days, except in cases of emergency. All operations, and

OPERATING ROOM.

all first dressings after operations, are done in the regular operating room. Suppurating wounds are dressed in another room. A few minor and simple dressings are done in the wards.

For operations, the surgeon has as assistants a house surgeon, an etherizer, and two nurses.

ARRANGEMENTS FOR THE OPERATING ROOM.

There are assigned, for every operation of importance, two nurses. The first nurse has charge of the handling of all the sterilized dressings, sterilized sponges, ligatures, and sutures. She is responsible for the cleansing of the tables, the asepsis of the instruments needed for the operation, and for their subsequent sterilization or cleansing.

To the second nurse is assigned the handling of such articles as are not thoroughly aseptic. She also renders whatever other service may be required, and, if desired, takes the place of the aseptic nurse. If so required, she prepares herself as directed in the accompanying rules.

The operating room is cleansed, washed, and scrubbed at stated intervals. The furniture, besides being cleansed as the room is cleaned, receives especial attention. The slate-topped tables are mechanically cleaned, scrubbed with hot water after operation, and before each operation are wiped with 1:1000 corrosive sublimate solution. The operating table is wiped with hot water, and the parts of the table near the operating field sterilized in the sterilizer for twenty minutes, after which the whole table is covered with a sterilized cloth until used.

The basins and jars of glass or agate ware, and trays,

are cleansed mechanically after each operation, rinsed clean with hot water; and before each operation, after being rinsed with cold water, are filled with 1:1000 corrosive sublimate. Those to contain dry material, such as dry dressings, are emptied just before the operation.

After operation, the instruments, except knives and needles, are rinsed in hot water, cleansed mechanically, and boiled for five minutes. They are then wiped with a sterilized towel, and placed on a shelf. Before each operation, they are placed in a cloth and boiled with bicarbonate of soda for ten minutes. Just before the operation, they are taken from the sterilizer and placed upon the table.

STERILIZATION OF KNIVES. — Knives, after being cleaned and sharpened, are placed each separate in a glass tube, the ends of which are plugged by cotton. The tubes are placed inside the dry sterilizer, and kept at a temperature of 130° C. for half an hour. They are then removed and placed in the instrument-case. Just before an operation, the tubes containing the required knives, and still plugged with cotton, are boiled for ten minutes in water containing bicarbonate of soda. Needles are treated in the same manner.

The dressings are of several kinds: —

DRESSING No. 1. — Aseptic dressing where no leakage is expected. A pad of dry sterile gauze (six layers), slightly larger than the field of operation, is placed on the wound; a second pad, of absorbent cotton, somewhat larger, is placed over this, and sheet-cotton is placed over the whole. A gauze bandage secures the dressing.

DRESSING No. 2. — Aseptic dressing where leakage is expected. This is the same as dressing No. 1, except that it is larger, and rubber tissue, or parchment paper, soaked in corrosive sublimate solution, is placed between the two outer layers.

DRESSING No. 3. — Aseptic dressing where a granulating surface is left uncovered. Same as dressing Nos. 1 and 2, except that a piece of sterilized rubber tissue is placed next to the wound.

DRESSING No. 4. — Antiseptic dressing. Similar to No. 1, except that next the wound a pad of gauze soaked in 1:2000 corrosive is placed; or where tubercular tissues are present iodoform gauze is used.

DRESSING No. 5.— Antiseptic and open dressings where much discharge is present. A cotton waste pad is used instead of the gauze pad. This is saturated with either of the following: Corrosive sublimate solution 1:2000; tincture of myrrh, thymol (1:5000), if much odor; or creolin (2%) solution. If there is much discharge, the whole dressing is covered with "macintosh" or parchment paper, unless the dressing is to be changed frequently, when no protective is necessary. Iodoform gauze is often used. Dressings by ointments are used as ordered.

PREPARATION OF TOWELS, SHEET, GOWNS FOR OPERATIONS, are as follows: —

When soiled, they are laundered. After this, they are folded up and placed in suitable bags. Before an operation, they are steamed for half an hour. The nurses', the assistants', and the surgeons' gowns, and the aprons are put on just before the operation, and only when the wearers have made themselves thoroughly clean.

Rubber aprons and tourniquets are scrubbed in soap and water, and rinsed in clean hot water. The aprons are then wiped off with 1:500 bi-chloride. Tourniquets and "Esmarchs" are put in 1:500 bi-chloride half an hour before the operation. This solution is weakened to 1:1000 just before operation.

PREPARATION OF THE PATIENT. — The patient is
bathed the night before operation. The field of operation is cleansed, first scrubbed with soap and water for
five minutes, then wiped off with ether freely, washed
with corrosive sublimate solution, 1:1000, for two minutes, and covered with a wet corrosive sublimate
dressing, which remains until the operation.

PREPARATION OF THE HANDS. — The surgeons', assistants', and nurses' hands are treated in the following
way : — In operations where special care is needed, and
where thorough preparation can be carried out, the
hands are soaked and washed in liquid German soap
and hot water for five or ten minutes, wiped dry with
a sterilized cloth, and then soaked for a minute in 1:1000
solution of corrosive sublimate.

THE DRESSINGS, AFTER OPERATION, are divided into
three classes, — the strictly aseptic, the ordinarily aseptic,
and the antiseptic.

The first class is always done in the operation room,
and with as great care as if it were an operation. This
is done under the direction of the surgeon, or according
to his direction. There is no change in these dressings,
except such as is prescribed by the surgeon, and the
strictest details are carried out in doing these dressings. They are only done at such times as are appointed
by the surgeon, and are not changed without his knowledge or consent. The dressings of this sort are the first
dressings after an operation, and the later dressings when
the case demands it. After the wound has been several
times dressed, it may be considered that there is no particular danger as far as infection is concerned ; but it still
requires some care. Under these circumstances, the
dressings are done by the house officer at such times as

may be convenient to him, and without the presence of the surgeon. No dressing is transferred from the aseptic of the first class, to the aseptic of the second or third class, without the knowledge of the surgeon. What are called the antiseptic dressings may be done by the nurse, or subnurse, without the presence of the surgeon. All dressings, cotton and gauze pads, bandages, and sponges, are cut of the requisite size, placed in copper cylinders, and sterilized by dry sterilization for one hour at 130° C. Before the operation, they are removed from the cylinders by the nurse with sterilized hands, wrapped in a towel, placed in a steam sterilizer and sterilized for twenty minutes, together with all towels or cloths that are needed for the operation. As far as possible, the dressings for each case are previously prepared, and sterilized just before the operation.

BRUSHES. — After their use, unclean brushes are thoroughly washed with soap and hot water, and then rinsed with clean water. They are placed in a jar filled with a corrosive sublimate solution (1:1000), where they are kept until again used. There are three brushes: one, which is reserved for the hands of the surgeon and assistants; one for the patient's skin near the wound; and one, to be used only in cases that are not clean. The scrubbing of the patient's skin directly after the removal of the antiseptic dressing and before the operation is done with pieces of gauze, either dry or wet, with 1:1000 corrosive sublimate.

STERILIZATION OF WATER FOR IRRIGATION. — The bottles to contain the water are rinsed out and then filled with water. They are then placed upon the boiling apparatus, which is half filled with water, and boiled for an hour. When this is done, a piece of sterilized cotton is placed in the mouths of the different bottles.

PREPARATION OF LIGATURES AND SUTURES : — Silk is cut in proper lengths and placed in the glass tubes, a definite number in each tube. These are wrapped in a cloth and steamed for twenty minutes. They are then put in a proper receptacle, for safe-keeping. Twenty minutes before an operation, one or more of the tubes are placed in the sterilizer, and resterilized with the cloths and dressings.

Catgut is also cut in proper lengths, and placed in glass tubes, a certain number in each tube. The tubes with the catgut are sterilized by being boiled in alcohol, and are kept in alcohol. In preparing the sutures for operation, sterilized catgut or silk is taken, the sterile needles are removed from the test tube where they are kept, and the needles are threaded and put into cloth, just before operation, and steam sterilized. As a rule, only a sufficient number of needles for one operation are threaded.

OUT-PATIENT SERVICE.

The growth of this department has been of interest; and the number of new cases applying for treatment in each of the past ten years has been as follows : —

1883	 350	1888	 723
1884	 344	1889	 843
1885	 389	1890	 986
1886	 528	1891	 992
1887	 646	1892	 1031

Previous to 1889, the work of the out-patient department was done in three small basement rooms of the hospital; but at that time the new out-patient building was opened, and has been in use ever since. It has proved to be a model of convenience.

The surgical out-patient department is open for the treatment of cases on four days weekly: two for the first out-service, and two for the second out-service; the first service having Mondays and Wednesdays, and the second service, Tuesdays and Saturdays. Patients can attend on these days between two and four in the

WAITING ROOM. — OUT-PATIENT DEPARTMENT.

afternoon, but are not admitted to the building later than this.

Treatment is free to any worthy person who applies, but apparatus is paid for in nearly every instance. There is no fund to provide for supplying apparatus. It has been found that it was a doubtful expedient to give apparatus free, as it was almost invariably neglected and carelessly treated under these circumstances.

Each out-service is under the charge of the surgeon.

His executive corps consists of the assistant surgeons,
two or three assistants, an externe, three to six dressers,
or clinical clerks, and two nurses. The assistants are
usually persons holding medical degrees, or hospital
graduates who have had special training and experience
in this class of work. They are invited to serve six
months or one year by the assistant surgeons, with the
approval of the surgeon, and are not appointed by the
hospital. They act as first assistants. The externe is
appointed by the Managers. He serves four months, and
then becomes an interne, his successor being appointed
in the same manner. The dressers and clinical clerks
are medical students, who are selected, from time to time,
by the assistant surgeon in charge. Their term of service is usually two to three months. The nurses are
members of the training school connected with the
hospital.

Each patient, at the first visit, is provided with a card
bearing his name and record reference. This card is

OUT-PATIENT DEPARTMENT. – SURGICAL.

TUESDAYS AND SATURDAYS, FROM 2 TO 4 O'CLOCK.

_____

Name ....

Out-Patient Records, Vol. ....*Page* ....

House Records, Vol. ....*Page* ....

_____

ALWAYS BRING THIS CARD WITH YOU.

brought at all subsequent visits. The patient is told to
return only on the days of the service on which he first
came. In this way, each patient remains continuously
in charge of the same surgeons.

When a patient is first brought for treatment, he goes
at once to the record room, where the history is taken.
He then receives the card previously mentioned, and returns to the waiting room, where he awaits his turn to be
assigned to one of the small examination rooms. There
he is seen by the assistant surgeon or his assistant, who

PLASTER ROOM.

examines the case, measures for apparatus if necessary,
and dictates the notes of the case to one of the dressers,
who afterwards enters them in the records. When the
surgeon has finished with the patient, and the treatment
is completed, the child is dressed and taken home. If
plaster-of-Paris work is required, the patient is transferred to the room especially fitted for this purpose.
When such work can be anticipated, the patient is sent
directly to this room. The assistant surgeon and his
assistants divide the work among themselves, and pass

from room to room examining the cases until all the patients have been seen. 6504 visits were made by surgical out-patients during the year 1892, making a daily average of over 30 cases. Inasmuch as during the winter the attendance is light, the clinics in spring and summer rise to figures greatly in excess of this daily average.

RECORDS. — The name of each new patient is entered in the record book and a card index of cases, and a card is given to each. The history is written out by the clinical clerks together, but the physical examination, diagnosis, and treatment are dictated by the assistant surgeon. At each subsequent visit, a record of the condition of the patient is made. These records are dictated to the clinical clerks by the assistant surgeon or his assistants. They are signed in every instance by the initials of the surgeon examining the case, no matter how often the patient is seen.

At the first visit, the history is taken upon a printed form, which has proved very satisfactory, and which is appended. In order to secure uniformity, the clinical clerk is instructed to fill every space, whether or not it seems necessary to him. In this way, fairly satisfactory records can be secured. With such a record, the surgeon can at a glance grasp the important facts of the case.

DATE. NAME.

FATHER'S NAME.

RECORDER. RESIDENCE.

BORN. FATHER'S OCCUPATION

DIAGNOSIS.

HEALTH OF FATHER AND MOTHER.
HEALTH OF BROTHERS AND SISTERS.
(Number Alive and Dead with Diseases.)
RELATIVES AS TO LUNG DISEASES.

9

RELATIVES AS TO RHEUMATISM.
"   "   JOINT DISEASE.
"   "   NERVOUS DISEASES.
REMARKS.

CONDITION OF PATIENT AS BABY.

BOTTLE FED OR NURSED.
PREVIOUS ILLNESSES (with Date of Exanthemata).

LOCATION OF PRESENT TROUBLE.
DURATION.
CAUSATION.
HISTORY OF INJURY, IF ANY.

LABOR.   (A detailed account of the labor should be given in
         all cases of Paralysis or Congenital Affections.)
FIRST SYMPTOMS NOTED.
PAIN, AND WHEN WORST.
         NIGHT CRIES.
LAMENESS, AND WHEN FIRST NOTED.
LOSS OF FLESH.
OTHER SYMPTOMS.

PREVIOUS TREATMENT.

REMARKS.

PHYSICAL EXAMINATION.

APPARATUS. — If appliances are necessary for any
case, the patient is told that such is the fact, and the
Sister or nurse who has charge of all financial matters
informs the parents what the cost will probably be. If
the patient makes a deposit with the Sister or nurse,
the measurements are taken and are sent to the shop;
but no measurements are taken until money matters
have been satisfactorily arranged. After a deposit has
been made, there is little chance that the patient will not
reappear and claim the apparatus when ready. Apparatus is furnished to the patients at cost. In rare instances only are patients unable to pay for it. The
assistant surgeon or his assistants have no financial deal-
ings with the patients. The Sister or nurse arranges

the amount of the deposit (generally about 50% of the whole cost), the terms of payment, gives receipts for money paid, and takes charge of the appliance orders, which she sends to the shop.

ADMISSION TO THE HOSPITAL. — If patients applying at the out-patient department are in need of operative or

GYMNASIUM, — OUT-PATIENT DEPARTMENT.

ward treatment, the matter is referred to the surgeon in charge, who fills out the following slip, and sends it to

RECOMMENDATION FOR ADMISSION.

Date...

Name...

Age.....................*Residence* ....

Diagnosis ....

Out-Patient Records, Vol. ... .*Page*...................

Surgeon ....

Not to be given to the Patient, but transmitted to the House Officer.

the superintendent, who admits the patient. This is not given to the parents of the patient, but is intended only for the information of the superintendent and visiting surgeon. This slip may inform the latter that the patient is referred from some other institution; that it is a case where a law-suit is involved; that the parents have been negligent; or that the case has done badly for some reason. The slip, in short, is a confidential communication from the assistant surgeon to the surgeon.

The discharge slip from the hospital is received by the out-patient department, where it is filed with the out-patient records of the individual case. By this system, the wards are kept in touch with the out-patient department, which is necessary in the long continued cases which form a part of any surgical clinic.

DISCHARGE.

Name..

Admitted..

Discharged...

Diagnosis ....

House Records, Vol.....................*Page*....................

Short Resumé of Treatment :

THIS SLIP TO BE GIVEN TO THE EXTERNE.

There is connected with the hospital a room for photography. In this way, much valuable data has been collected.

In the basement is the museum. This contains specimens of the different types of apparatus used in the treatment of orthopedic cases and pathological specimens. The apparatus has been collected from time to time, and is a valuable collection for the instruction of workmen and students.

In regard to the treatment of especial classes of cases, the following details may be of interest: —

HIP-DISEASE. — Cases of hip-disease are directed to come to the hospital every three weeks if acute, or less often if the assistant surgeon sees fit. The plaster extension is changed at the hospital, except in a few cases (by the mother) which live at a long distance, where frequent visits are impossible. Measurements of all cases of hip-disease are taken at every second or third visit, and, to secure uniformity, the following blank is generally used, from which the data is transferred to the record book.

RECORD OF HIP MEASUREMENTS.

Name.. *Date*............................

Out-Patient Records, Vol...................*Page*................

Length from Ant. Sup. Spines . . . . . . *R*....................*L*...................

{ * *Length from Umbilicus* . . . . . . *R*....................*L*............. }

{ * *Distance between Spines* . . . . . . ————————. }

     Adduction..................*Abduction* ....

     Permanent Flexion ....

Circumference of Thigh . . . . *R*....................*L*...............

Circumference of Calf . . . . *R*....................*L*...............

Amount of motion: —

    In Flexion ....

    Abduction ....

    Rotation ....   *Temperature*................

Remarks..

---

* *After the amount of abduction or adduction has been calculated, it should be put down in degrees and these measurements crossed out.*

To insure proper care at home, patients with hip-disease are provided with the following slip of directions when the hip-splint is first applied.

DIRECTIONS FOR THE CARE OF APPARATUS.

Read this carefully until you have thoroughly learned it.

Taylor Hip Splint. — Apparatus to be worn constantly. If the child has pain, return to the Hospital. Always put the splint on with the child lying flat on its back. Put the splint in place, and then tighten the band in the groin and buckle it fast; put the extension straps on the little pegs at the bottom of the splint, and wind with the key until it is as tight as the child can bear it; then buckle the strap around the calf of the leg. The bottom of the foot should never touch the bottom of the splint. Keep the bands in the groin buckled tight, and tighten the splint with the key as often as the straps get loose. When the splint has to be removed for any reason, have some one pull down on the leg gently all the time until the splint is replaced. Bathe the groin and powder it daily, and watch for chafing. Do not take the splint off oftener than is absolutely necessary, and never leave it off. Have the child lie down at least two hours each day. Unless you are told to the contrary, come back in at least three weeks, and sooner if the plaster gets loose. In loosening splint, unwind with the key, not by springing the catch. Keep extension straps tight. If the child has pain or night-cries, return at once to the Hospital. Return at once if repairs are needed on the splint.

SPINAL CARIES. — At the first visit, a cardboard tracing of the spine is taken, as shown in the report on "Caries of the Spine," and filed as a record. Patients are directed to come every month, unless they live at a long distance; and at every second or third visit, a new tracing is taken, dated, and is tied up with the former ones. In acute cases, tracings are taken more often. The tracings are hung on hooks in the record-room, and are arranged alphabetically so as to be accessible. Patients are supplied with the following directions at the first visit, where a back-brace is applied.

DIRECTIONS FOR THE CARE OF APPARATUS.

Read this carefully until you have thoroughly learned it.

Back Brace. — Apparatus to be worn constantly. Brace to be removed twice a day, as follows : Child lying face down, unbuckle apron, remove brace, wash back, dry thoroughly, powder with "baby powder;" replace brace by buckling the bottom buckles first, then the next above, and so on, fastening the shoulder-straps last. When ready, the apron should be perfectly smooth, and the brace tightly buckled to the back. Never allow the child to sit up until the brace is properly applied. Always examine the back carefully while washing, and if any red spots are seen, report at once at the Hospital. Examine the brace several times daily to see that it is always kept firmly applied.

A child should return to the Hospital as often as possible until the apparatus is properly fitted; then, unless told to the contrary, once every four weeks. If the child is not perfectly comfortable, return at once.

KNOCK-KNEE AND BOW-LEGS. — Tracings on common wrapping-paper are taken by a pencil held upright and made to follow the outline of the leg. These are made at intervals of three months, and are kept in portfolios (being arranged alphabetically) in the museum.

CLUB-FOOT. — Records of feet are kept by photographs and plaster-of-Paris casts. Patients wearing club-foot apparatus are supplied with the following directions.

DIRECTIONS FOR THE CARE OF APPARATUS.

Read this carefully until you have thoroughly learned it.

Club-Foot Shoe. — Apparatus to be worn constantly. To be removed twice daily, foot to be washed, dried thoroughly, powdered with "baby powder." Reapply apparatus as follows: Turn down the upright, buckle foot firmly to the foot-piece; see that the heel is well applied, resting on the sole-plate; if not, reapply. Turn up the upright, and fasten to the calf of the leg. Always examine the foot carefully while washing; if any red spots are seen, return at once to the Hospital. Apparatus to be worn inside the shoe. Examine the apparatus several times daily to see that it is properly applied. Child to return to the Hospital as often as possible until the apparatus is properly fitted; then, unless told to the contrary, once every four weeks. If the child is not perfectly comfortable, return at once.

THE SURGICAL APPLIANCE SHOP.

The appliance shop has been in operation for eight years, and during that time has increased in size from the employment of one workman to that of six men and one woman. The shop is under the immediate control of the surgeons connected with the hospital. All matters of discipline, however, and of the details

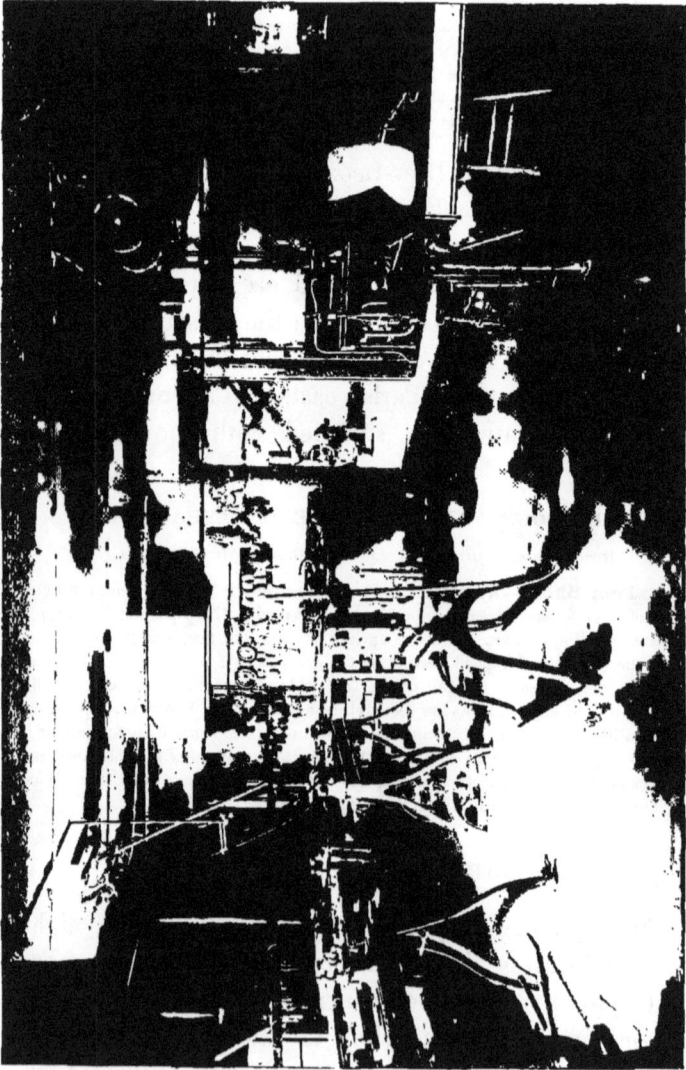

SURGICAL APPLIANCE SHOP.

of the shop, such as the hiring and discharging of men and the details of distribution of the work in the shop, are left to the foreman. Apparatus is furnished to patients at the cost of manufacture, and is only manufactured after the delivery of a signed order of an officer of the hospital, who becomes responsible for the payment. All work on the apparatus is done at the shop. The working force consists of a foreman, one forger, two men employed on finishing work, two boys, and one seamstress.

A part of the lower floor in the out-patient department at the hospital is used for the appliance shop. It contains four rooms, — one small room, which is used for the forging, and which contains two forges and two anvils; and a large central room, which is used for the finishing work, and which is fitted up with benches, vises, and lathes, in this room also is placed a five-horse power Otto gas-engine, which furnishes motive power; a third small room, which is used entirely for the grinding and polishing; and a fourth room, for the leather work. The heavy leather work on the splints is done by the workmen in the shop; the lighter by the woman, who has her space in the museum, where is also done any ordinary sewing which is found to be necessary.

The method of ordering is indicated by the accompanying forms.

For orders, blanks are bound in book-form, on which are spaces for the date, name of the person who is to be responsible for the payment of the order, name of the patient, and a description and measurements of the apparatus. This is attached to a stub, which is left for reference and the convenience of the person ordering

.............189...

Name of Patient:

...................................189....

This space is for signature of party responsible for the payment of this order.

Apparatus: *Name of Patient:*

Apparatus:

Measured by   *Measured by*

This blank is then sent to the workman, who places it upon his day order-book, which has on its margin

No......   *No.*

Date Received,..................... *Date Received,*......

Ordered by

Cost,

Name of Patient:

Extra,

Apparatus:

Total,

Cost,.....................

Date Delivered,..................... *Date Delivered,*...

blanks as seen in the illustration, on which he fills out the date at which it is received and delivered, and the cost of the apparatus. As the apparatus is finished, this slip is torn from the book and put on file, and, at the end of the month, is sent in, and the account rendered to the different individuals. This book also contains a stub, which gives only the date, and the name of the apparatus, which remains in the shop in case of necessity of reference.

There are two kinds of apparatus made in the shop. First, that of some recognized pattern; and second, any new apparatus or experimental work which may be carried on by any of the surgeons connected with the hospital. For the first, an order is made which simply states the kind of apparatus, and gives certain measurements which have been determined upon. This is then given to the foreman, and the details of the work are taken in charge by him, and, when finished, returned to the person ordering. As most of the apparatus which is manufactured at the shop is mainly of recognized form, in measuring the name of the apparatus only is given, with the necessary measurements. Some patterns of the different sizes and shapes of certain parts are kept, and are used in place of measurements. These are then sent to the shop, and duplicate parts of the splint are made from them.

PART II.

POTT'S DISEASE.

SPINAL caries forms $26\frac{8}{10}\%$ of the total number of patients which have been treated at the hospital. The flexibility of the spine is one of the principal tests used to establish the diagnosis of caries of the

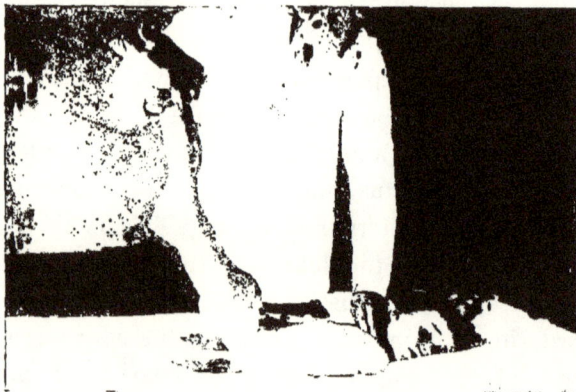

NORMAL FLEXIBILITY OF SPINE.

spine. (The above cut shows the flexibility of a normal spine; that on page 141, the stiffness of a spine in low dorsal caries.) The treatment, as carried out in the hospital, may be classified into constitutional and surgical.

CONSTITUTIONAL TREATMENT. — This is applicable to all cases, and consists in improving the patient's general physical condition, so far as the local conditions of the spine do not contra-indicate, by careful attention to hygiene and diet. The patient is treated principally in the out-patient department, but is taken into the wards whenever it is necessary, and is exceptionally sent to the Convalescent Home, at Wellesley. Medicines are administered as they are indicated, — such as cod-liver oil, iron preparations, and the hypophosphites. Other medicines have been tried, but without marked benefit.

SURGICAL TREATMENT. — The essential part of the treatment of spinal caries is antero-posterior [1] fixation of the spine, with leverage so applied as to place and sup-

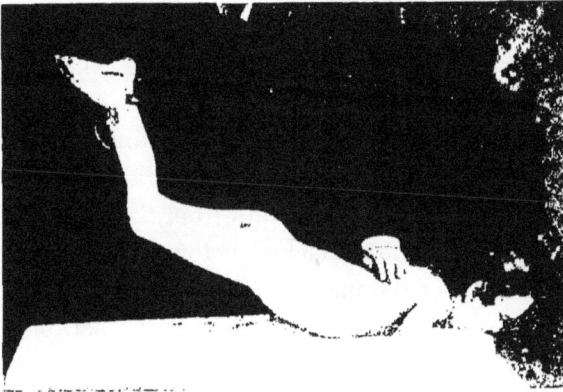

RIGIDITY OF SPINE IN POTT'S DISEASE.

port the super-incumbent pressure upon the transverse processes, and this is carried out, with more or less completeness, until the patient is cured. Apparatus is modified, abscesses are opened, and extension and counter-extension in bed are applied in special cases. The means for securing antero-posterior fixation of the spine and support of the super-incumbent pressure, and for placing that pressure upon the transverse processes, are either —

_____

[1] According to the new anatomical nomenclature, this should be "dorsalventral."

1. *Bed-treatment,* or
2. *Ambulatory protective treatment.*

1. BED TREATMENT. — The beds which are used in the hospital wards are of uniform design and size. They are made of iron, and are finished with " Japan," — the only ornamentation consisting of four small brass balls, which are placed on the top of the four corner posts.

The bed is sixty-two inches long in its extreme length, and thirty-three inches broad, while the extreme height is forty inches. The mattress rests upon a coarse lattice made of thin strips of iron, which is so situated that the top of the mattress, or the position of the child, is about two feet from the floor.

The bed is supported upon four round posts, which are placed at the corners. These posts are connected at the top by small round bars, and from these are perpendicular bars which are attached below to bars placed at about the level of the mattress. This frame-work at the sides can be lowered at will, making an open bed or a crib, as is desired.

The child is placed in bed upon an oblong bed-frame made of gas-pipe. (Vide description under hip-disease.) This frame is modified at times by curving it upwards opposite the kyphosis, so that it brings pressure to bear on the site of deformity, thus preventing the sagging of the spine which at times occurs when the patient is simply resting upon a mattress. (Vide page 143.)

Bed-treatment is given to many cases in the course of the disease. The antero-posterior back-brace is combined with bed-treatment, it being necessary at times, in very sensitive spines, to use every means of fixation that we possess. The rule is, however, to keep the patient in bed as little as is consistent with its spine being protected against super-incumbent pressure or

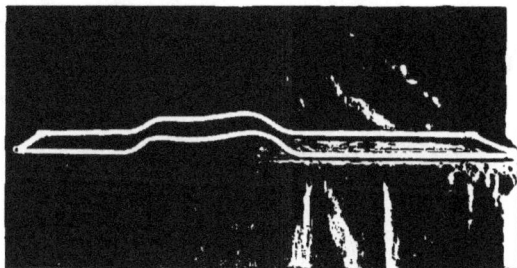

SELVA WIRE FRAME.

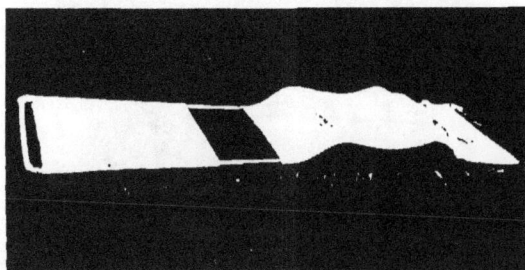

SELVA WIRE FRAME COVERED.

SELVA WIRE FRAME APPLIED.

motion. Infants are treated on a frame, with additional fixation secured by a back-brace, or plaster-of-Paris jacket.

Where the child is not old enough to walk, or where there exists concomitant hip or knee joint disease, it is kept on a frame which permits the spine to be fixed, and

TAYLOR BACK-BRACE WITH HEAD SUPPORT.

yet enables the parents to move the child about on the frame at their will.

2. AMBULATORY PROTECTIVE TREATMENT. — The antero-posterior back-brace which is commonly used in the hospital is a modification of Taylor's, and is made as follows: —

A tracing is taken of the vertebral outline of the back, over the transverse processes, with a lead or block-tin strip, while the patient lies prone upon a flat, unyielding surface, the arms being extended by the side. This outline is traced on a sheet of press-board and then cut out, and forms an accurate profile of the spinal column, from as high up as is possible in the cervical region to the sacrum.

This is used as a guide to the workmen in curving the

METHOD OF MEASURING DEFORMITY IN POTT'S DISEASE.

uprights, and in determining the length of the brace. The distance of the uprights apart is the distance between the transverse processes. Cross-bars, usually two, are placed between these uprights, at appropriate intervals, to insure firmness. The apparatus is fixed to the patient by being buckled to a cloth-apron well fitted to the anterior surface of the body, and extended from the supra-sternal notch to the pubes. Two plates are attached to these uprights at a point to rest against the lower incline of the kyphosis, especially if the boss is marked. The point of adjustment of these two plates

10

should be between the spinous and transverse processes of the vertebræ, and, by their careful adjustment and management, we find that the back-brace can be made very efficient. This apparatus is modified to meet complications and special cases, combined with head-supports, and is the standard method of treatment.

TAYLOR BACK-BRACE.

Plaster-of-Paris jackets are frequently used as a substitute for the antero-posterior back-brace under the following conditions: — In infants and children whose parents cannot be taught how to adjust the back-brace; occasionally in the convalescent treatment of the disease; and often for the sake of economy. There are, however, a few cases of low dorsal caries which experience shows are better supported by a plaster-of-Paris jacket than by any other means.

There are two methods used in applying plaster-of-Paris jackets, — one is partial suspension in an erect attitude; the other, recumbency. During the application of all jackets, the patient is put into that position which is found in his case will best straighten the spine and remove pressure from the diseased portions.

In the first method (partial suspension in an erect attitude), a head sling is used to steady the patient and to remove the weight of the upper part of the trunk, and the hands are made to grasp either a bar above the head or the return rope of the head sling. Enough traction is made to straighten the spine as much as desired, but no attempt is made at forcible straightening by complete suspension.

In the recumbency method of applying a jacket, the patient is laid prone, with the arms above the head, on a hammock, which consists of a stout cloth a little longer

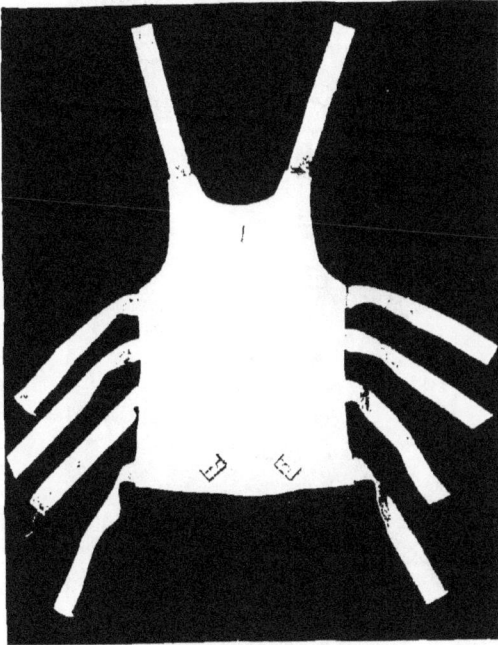

APRON FOR TAYLOR BACK-BRACE.

and wider than the child, stretched over the ends of a rectangular gas-pipe frame. One end of this cloth is secured to the upper end of the frame; the other end of the cloth is attached to a movable bar which is connected by a screw to the lower end of the frame. By means of this screw, the tension on the cloth may be regulated. To apply the jacket, the cloth is slit closely along the

sides of the child, and the bandage in the application
passes through these openings, thus incorporating in the
jacket that part of the hammock on which the child lies.
As the position causes pressure on the abdomen, it is
usually necessary to insert padding beneath the under-
vest, which serves the purpose of a "dinner-pad." In

TAYLOR BACK-BRACE APPLIED, SHOWING CHEST PIECE.

the other method of applying the jacket, pads are placed
over the transverse processes along the lower incline of
the kyphosis, both for the pressure exerted and for protection of the prominence from chafing. Below, the
jacket is made as long as can be worn when the child
sits down, and, above, is carried to the level of the junction of the second rib and sternum, and cut out in the
axillæ so that the shoulders are not much raised.

The hammock method (page 150) is applicable to all cases except those in which it is desirable to keep a decided lordosis. The special advantages are, ease to the patient, particularly in feeble children, and the ability to regulate the amount of extension of the spine by the tension of the hammock.

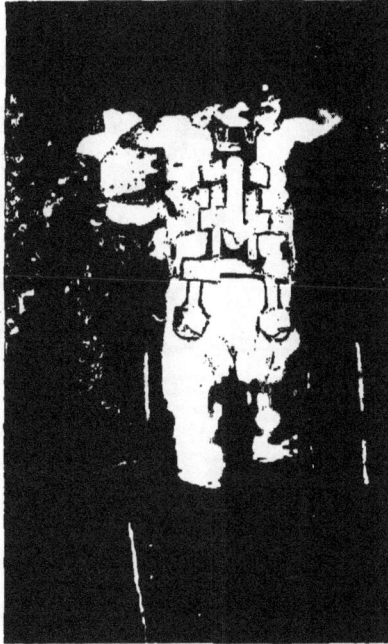

TAYLOR BACK-BRACE, SHOWING BACK.

Corsets of leather, paper, etc., are often used during the convalescent treatment of the disease.

The treatment varies according to the regions involved, and these modifications will be described under the following headings: —

1. Cervical caries.

2. High dorsal caries, disease at or above the fifth dorsal vertebra.

3. Dorsal caries, disease below the fifth dorsal vertebra to the eleventh or twelfth dorsal.

4. Lumbar caries.

5. Complications.

1. *Cervical Caries.* — When a child presents itself
with cervical caries, with the disease acute and the cervical rigidity marked, it is treated by rest in bed with
a head-extension. A light weight (from one half-pound
to three pounds) is attached to a cord which passes over

METHOD OF APPLYING A PLASTER JACKET.

a pulley at the head of the bed, and the head of the bed
is raised to give counter-extension by the weight of the
body (page 152). Not infrequently, twelve hours of this
treatment will render a child perfectly comfortable. As
soon as the pain and spasm of the muscles have subsided,
a more permanent form of apparatus is adjusted. This
consists of a Thomas collar in the high cervical caries,
and an antero-posterior back-brace and head-rest in low
cervical caries.

In cases of cervical caries where the Thomas collar
will not support the vertebræ, an antero-posterior back-
brace and head-rest are used. A number of forms of

head-rests have been tried, but the simple ovoid ring of Taylor has been found the most satisfactory. The illustrations on pages 153, 154, and 155 show the various forms of apparatus applied.

A head-rest devised by Dr. J. E. Goldthwait is used

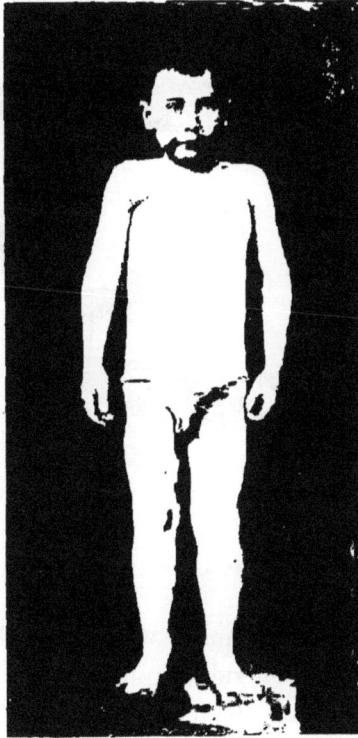

PLASTER JACKET.

in the treatment of cervical caries where it is in the lower cervical region, and the head is carried by the child well forward. The Goldthwait collar consists of two flat steel strips which extend from the lumbar region along the side of the spine over the shoulder down on the anterior part of the chest to the lower third of the sternum, and joined at this point with a piece from the other side.

At the highest point of the shoulder, opposite the anterior edge of the trapezius muscle, is attached a wire which extends upward as far as the level of the angle of the jaw, and at this point is turned at a right angle and extends forward horizontally as far as the tip of the chin,

TRACTION IN CERVICAL CARIES.

where it is joined by a similar wire from the other side. The wires from either side support a concave plate of hard rubber in which the chin rests. The two vertical portions of the wire behind are connected by a small flat band of steel, which is hinged to one side and is secured to the other by a catch. From the centre of this extends upwards a small wire to the level of the occiput, and at the top of this is fastened a small plate of brass which makes forward pressure on the back of the head. Another form of head-rest, that has been found useful and inexpensive, is shown on page 154.

In that type of disease which commences with slight symptoms or occasional attacks of cervical rigidity, patients are treated by the immediate application of a Thomas collar, or an antero-posterior back-brace and head-rest. When, at any time in the course of the dis-

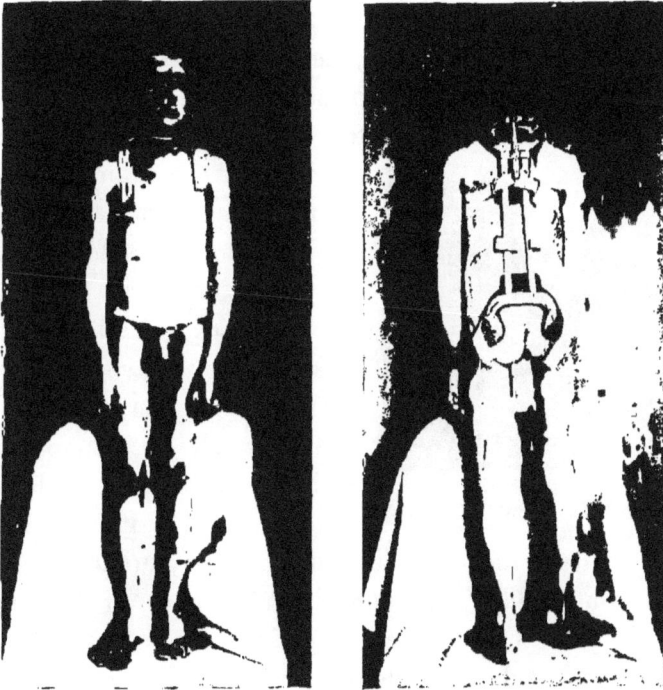

TAYLOR BACK-BRACE WITH HEAD SUPPORT, APPLIED.

ease, the patient develops grunting respiration, especially with dyspnœa, or rests the chin upon the palm of the hand, or complains of severe pain, which is not relieved by an efficient apparatus, or is losing weight or appetite, he is put to bed on a frame and kept there until relieved of his urgent symptoms. The majority of cases of cervical caries do not require bed-treatment, but are treated in the out-service with satisfactory results. The differential diagnosis of cervical caries from spasmodic torticollis is only made at times by the treatment of the case.

2. *High Dorsal Caries, Disease at or above the Fifth Dorsal Vertebra.*—The spine above the fifth dorsal vertebra is so movable that an ordinary antero-posterior back-brace does not secure it, and, unless the head and neck are supported as in cervical caries, the diseased area is not kept at rest; so that all cases of spinal caries above the fifth dorsal vertebra are treated like cervical

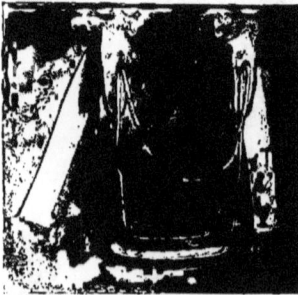

BENT WIRE HEAD SUPPORT.

caries by re-enforcing the spinal brace with a head-rest which secures the upper part of the vertebral column. Acute cases, or those with exacerbations, have bed-treatment. When the acute cases are relieved, an antero-posterior back-brace is ordered, and, as soon as it is satisfactorily fitted, the child is allowed to sit up for one or two hours daily. During the remainder of the time it is in bed on the frame. The time of being up is gradually increased as the child grows stronger, until it is up and about during the entire day. At night it sleeps on the frame.

Tracings of the spine are taken when the child first comes to the hospital, and at regular intervals (once a month) afterwards, and if there is any increase in the kyphosis, it is treated by rest in bed again, with the curved frame or pads, until there has been improvement.

In the out-patient department, the patient reports for treatment every two or three weeks, and its weight is recorded, and its general condition watched. Pain or any marked loss of flesh, elevation of temperature, or resistance in either of the iliac fossæ, suggest an extension of the carious process or the formation of an abscess, and it is admitted to the hospital, and treated by rest in bed

TAYLOR BACK-BRACE APPLIED.　　PRESSURE MARKS FROM TAYLOR BACK-BRACE.

until the symptoms are relieved. If the back-brace is carefully applied by the parents and well looked out for, the cases do well and the deformity should not increase; in fact, the leverage which is exerted by accurately adjusted padded plates holds the deformity while the child grows, and the lengthening of its spine makes the deformity much less apparent.

3. *Dorsal Caries, extending from the Fifth to the Eleventh or Twelfth Dorsal Vertebra.* — This is the common seat of the disease. The antero-posterior backbrace is used in the treatment of the disease in this location, and the treatment is the same as in the high dorsal caries, except that head-supports are not needed. The accurate adjustment of the padded plates in dorsal caries is of great importance, as by them the kyphosis can be controlled. The amount of pressure that these plates can exert is shown in the illustration on. page 155. Where it tends to increase in spite of the accurate adjustment of apparatus, the patient is placed in bed on a frame, and pressure is brought to bear against the boss.

4. *Lumbar Caries.* — When the lumbar vertebræ are involved, the patient is given bed-treatment, with pads applied to the seat of the disease a greater length of time than with dorsal caries. Since the bodies of the vertebræ normally curve forward, and are larger in the lumbar region than in the dorsal, there is less of a boss seen at this point, and, occasionally, fixation of the spine with the patient in an hyper-extended position, is the only local objective symptom found. In adjusting the back-brace or plaster-of-Paris jacket in these cases, care is taken to relieve the diseased bodies by throwing the superincumbent weight upon the transverse processes, and by preventing the spine from bending forwards.

In applying a plaster-of-Paris jacket, the patient stands erect, reaches upwards and grasps a bar above his head, and then sways slightly forwards in order that the weight shall be removed from the intra-vertebral discs or bodies of the vertebræ. These cases usually do well; but if the intestinal symptoms, and referred pains through the lumbar plexus of nerves are not relieved by apparatus, the patient is given bed-treatment, with careful attention to increasing the anterior lumbar curve by pads, and extension and counter-extension is applied to the spine.

5. *Complications.* — There are a number of complications which may arise in the treatment of spinal caries. The secondary changes of marasmus are treated according to the usual methods, and, when the case is infected

AUG.'88.   DEC.'88.   AUG.'89.   JULY.'90.

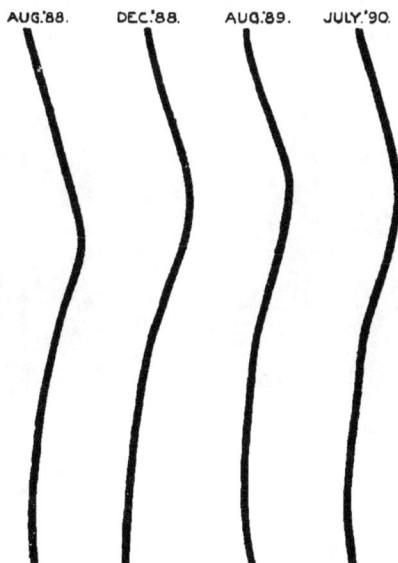

RECESSION OF THE DEFORMITY IN A CASE OF
POTT'S DISEASE.

with general tuberculosis or meningitis, the treatment is symptomatic, palliative, and protective.

CORRECTION OF DEFORMITIES. — Kyphosis of course exists to a greater or less degree in all cases, and treatment can rarely do more than prevent its increase. Occasionally the boss can be somewhat diminished, and these results are accomplished, first, by rest in bed, combined with the use of pad-pressure; and second, leverage, as

obtained by the careful and accurate adjustment of plates,
as shown on page 157.   Lateral deviation of the spine,
especially in the early stages, is treated by rest in bed,
with extension and counter-extension; but by attention
to the detail of fitting the Taylor back-brace and an
accurately fitting neck-piece, the deviation can usually
be prevented.

PSOAS CONTRACTION. — This complication is treated by
rest in bed, and traction in the line of deformity.   Day

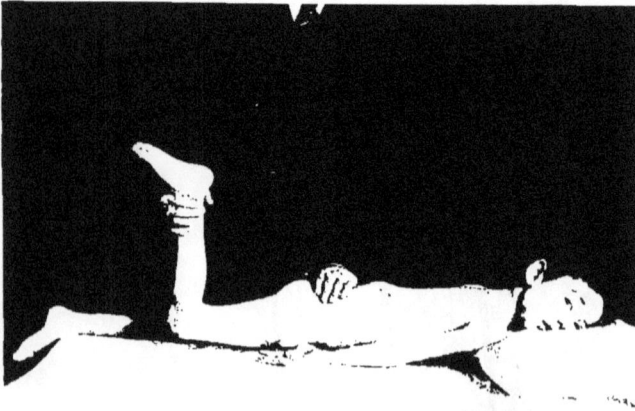

EXAMINATION FOR PSOAS CONTRACTION.

by day, the inclined plane upon which the limb rests is
lowered, and, in a varying length of time, the deformity
can be corrected.   When the deformity is dependent
upon an abscess, the abscess is evacuated, and the deformity quickly yields to traction.   Its presence is
detected by the test shown in the above illustration.

ABSCESSES. — Retro-pharyngeal abscesses are rare, and
are treated by an incision in the median line of the
pharynx.   In one instance, the abscess was evacuated
satisfactorily through an incision just behind the upper

part of the sterno-cleido-mastoid, evacuating the abscess through the side of the neck.

Psoas abscesses are of three forms : First, those which are acute, and which appear early in the disease and are grave in consequences. In this class of cases, the child is placed on a frame (that it may be moved into the sunlight and fresh air day by day), and kept there until the symptoms subside. When the abscess forms a distinct tumor in the iliac fossa, and, in spite of tentative measures, increases in size, or where the child's general condition is failing, it is thoroughly evacuated in a manner described below.

The second class of case is where the abscess appears at about the end of a year. Unless it diminishes in size, and the patient's general condition grows better, an operation is performed.

The third class of case is where the abscess is simply the detritus of the carious process. By careful attention to hygienic conditions, and with rest in bed and complete fixation of the spine by apparatus, absorption not infrequently occurs. Aspiration is rarely used; and if the abscess is not absorbed, it is evacuated.

When abscesses are opened, they are evacuated thoroughly. An incision is made at the outer border of the quadratus lumborum, and the channel of pus is tapped as near the spine as possible. Other openings are made in the iliac fossa, or on the anterior surface of the thigh, in order that the abscess may be completely evacuated and its "tubercular lining" removed. Drainage-tubes are used, and the drainage is thorough. In interfering with psoas abscesses, we have come to feel confidence that if we drain thoroughly and antiseptically, there is very little danger. The drainage-tube is retained in position for a long time, just so long as it is suspected that there is any tubercular material to be drained away. Iodoform (glycerin emulsion), or peroxide of hydrogen, in solution, are used for injecting the sinuses; but the main dependence is placed upon the primary operation for the thorough evacuation of the pus and tubercular material. Not infrequently we find a few crumbs or fragments of bones at the bottom of a psoas cavity, which are removed; but a formal removal of the bodies of the vertebræ has not been performed.

The dressings which are applied in these cases are the following: — Gauze soaked in sublimate, 1 : 5000, for in all cavities great care is taken about the strength of the sublimate that is used. Over this are placed pads made of absorbent cotton and gauze, which are sterilized by steam, and which are held in position by baked sheet-wadding and sterilized gauze bandages. Dressings are as infrequent as is possible. Temperature, odor, or saturation of bandages with discharge are the indications for redressing the wound. As soon as possible, a back-brace is adjusted, or a plaster-of-Paris jacket is fitted, and the child is placed on its feet and allowed to be about.

PARAPLEGIA. — The most common and important complication seen in which the nervous system is involved is paraplegia. It is considered of the utmost importance to stop the process causing irritation or pressure as soon as possible, and whenever, in the course of the disease and often when the child is properly protected by a well-fitting antero-posterior back-brace, there develops increased tendon-reflexes, stumbling in the gait, paresis, and paraplegia, the treatment is admission to the hospital, bed-treatment, with extension to both the lower extremities, and counter-extension by raising the foot of the bed.

Iodide of potash, in increasing doses, is usually combined with rest, extension, and counter-extension. In varying lengths of time (from a few days to a few months) this treatment is followed by a gradual return of sensation and motion, and, as soon as sensation returns, electricity and massage are systematically applied to the paralyzed limbs. Only once has it been found necessary to perform laminectomy for the relief of pressure-paralysis. In this case the paraplegia had existed for fourteen months, the child having been under bed-treatment, with extension and counter-extension, for six months with no improvement. The laminæ of three vertebræ were removed, and an abscess was found in the spinal canal and posterior mediastinum, which pressed on the dura. The sensation in the limbs returned within three hours. The patient was walking about in forty-two days, and now, two years after the operation, is well, with the sinus closed, wearing an antero-posterior back-brace for protection.

Cases of spinal caries are treated in the foregoing manner. The duration of treatment is usually four to five years. When twelve to eighteen months have elapsed since the disappearance of all evidence of an active carious process, and consolidation is beginning, the patient is considered convalescent, and the last stage of treatment is begun. The back-brace is omitted for short periods of time, and the child is kept under observation. If the case does well, the apparatus is gradually discontinued. Any failure in its general condition, increase in deformity or pain, is an indication for the immediate renewal of active treatment. This supervision is kept up until after puberty; and whenever the child is liable to receive injuries from its mode of life, the spine is still protected. A lighter form of steel support is often kept on until the child becomes an adult.

H. L. B.
H. W. C.

HIP-DISEASE.

THE methods of treatment of hip-disease employed at the hospital may be classified as follows:

a. Bed-treatment, for acute symptoms, and for correction of deformity.

b. Ambulatory treatment, with protection, with or without traction.

c. Operative treatment (incision, excision, or amputation).

BED FRAME, BRADFORD'S.

a. BED-TREATMENT. — Patients suffering from acute symptoms, or cases with symptoms threatening acute inflammation, are kept under bed-treatment until the alarming symptoms have subsided. Traction is always used in these cases, either by weight and pulley or by means of traction appliances. The child is placed on a fixation-frame so that he can be lifted without jar to the joint. The frame in use at the hospital has the advantages of being cheap, is readily made, and permits adjustment to the varying positions of abduction and adduction met with in different patients and in different phases of the disease. This frame is oblong in shape, and is made by inserting four pieces of cut gaspipe, one half inch in diameter (or larger in the case of heavy patients), into four rectangular gas-fitter's joints. The frame is two inches longer than the child, and it is as wide as the distance from the outer surface of one shoulder to that of the other. A stout sheeting covers this frame, and is laced in the back, where a space is left open

PATIENT ON FIXATION FRAME FOR CORRECTION OF DEFORMITY.

beneath the patient's buttock. Webbing-straps with buckles pass over the patient's trunk and under the frame, crossing the shoulder on one side and the axilla on the other. The child is placed upon this frame, a towel is placed over the hips and around the frame, and traction is applied either by weight and pulley or by means of a traction-splint. If necessary, the frame is fastened to the sides of the bed in the case of very restless children. In addition to the frame, sandbags can be placed at each side of the affected limb, to prevent any lateral motion of the leg. This, however, is ordinarily not needed. When the child is placed upon the frame and a traction-splint is applied, the patient can be moved from room to room without danger of injury to the hip. It is essential that, in addition to traction, counter-trac-

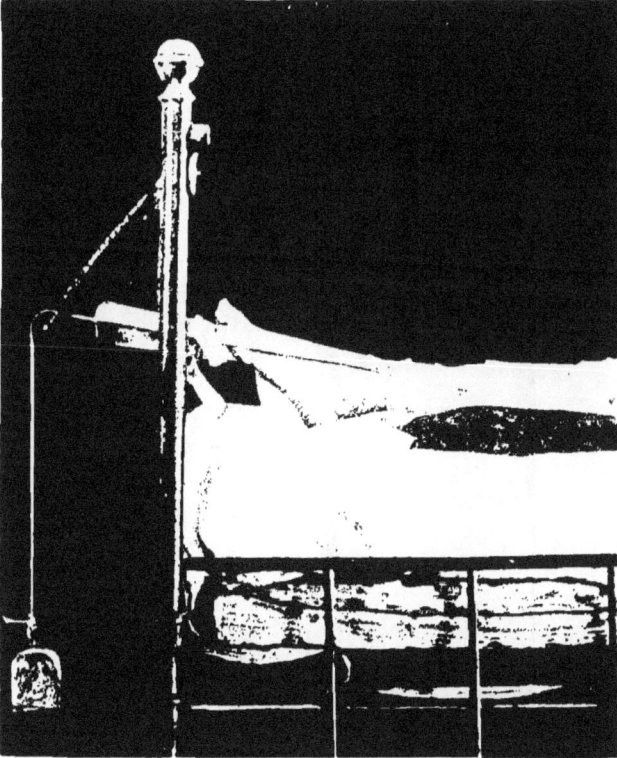

FOOT-PIECE FOR WEIGHT AND PULLEY TRACTION.
[*From the Fiske Prize Fund Essay.*]

tion be applied. This can be done, if the weight and pulley are used, by raising the lower end of the bed by blocks from four to six inches high. Where traction-splints are used, counter-traction is furnished by perineal straps. The traction-splint used, and the methods of correcting the deformities, will be described later.

Bed-treatment is considered preferable to ambulatory treatment.

1. When sensitiveness of the hip is present, as manifested by night-cries, or sensitiveness on moving the limb.

PLASTER-OF-PARIS SPICA BANDAGE.
[From the Fiske Prize Fund Essay.]

2. When deformity is present to a marked degree.

3. When abscess is present.

4. In double hip-disease, until the stage of convalescence has been well established.

But the choice between bed and ambulatory treatment is necessarily a matter of judgment, varying in individual cases. Methods of partial fixation, such as the employment of the well-known Thomas splint, the plaster of-Paris spica, or traction by means of a splint embracing the thorax, have all been occasionally used; but they have not been found satisfactory in cases where acute

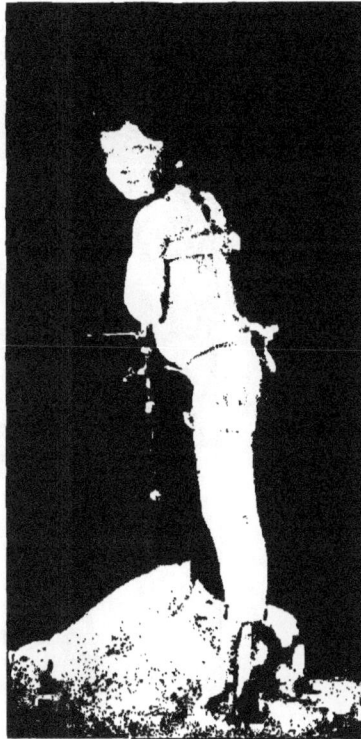

TRACTION SPLINT EMBRACING THE THORAX.
[*From the Fiske Prize Fund Essay.*]

symptoms demanded careful fixation, and they have seemed unnecessary in the convalescence of sub-acute cases when locomotion is permitted. For this reason, the long traction-splint has been used in preference, not in the opinion that it furnished fixation of the hip-joint, but inasmuch as locomotion is only permitted when absolute

fixation is no longer necessary, and because (as will be
seen in an accompanying paper on traction) the experience in the work at the hospital points conclusively to
the advantage of continual traction in hip-disease (as
has been claimed by Drs. Taylor, Sayre, and Shaffer)

LONG TRACTION SPLINT.
[*From the Fiske Prize Fund Essay.*]

for a period much longer than is demanded for absolute
fixation.   Lateral traction has of late been somewhat
used in painful cases.

b. AMBULATORY TREATMENT. — Patients suffering from
hip-disease should, as far as possible, be given the benefit
of exercise, provided the hip is protected in such a way

that no injury to the joint can take place. Apparatus for ambulatory treatment should —

1. Protect the diseased joint.

2. Furnish traction, and diminish the muscular spasm from crowding the femur into the acetabulum.

3. Limit the amount of motion.

It is manifest that it is impossible to fix the joint absolutely, and to allow the patient at the same time to

THE APPLICATION OF LATERAL TRACTION DURING RECUMBENCY.

walk about. The amount of motion allowed depends upon the condition of the patient's joint.

The ambulatory apparatus ordinarily used at the hospital is in most cases the long traction appliance with, in addition, crutches and a raised boot on the well limb. The instrument in general use at the hospital is a modification of the Davis-Taylor apparatus, which, for cheapness, is furnished with a windlass traction-attachment, instead of a ratchet and pinion extension. The object of traction being to overcome muscular spasm, a substantial amount of traction is needed. Traction made

from the dorsum of the foot and the heel is inadequate,
as pain is caused by any considerable pressure in that
region. The best pull is that furnished by adhesive-
plaster straps. One strap is applied longitudinally along
the side of the leg, while another strap fastened to this

LONG TRACTION SPLINT WITH AND WITHOUT CRUTCHES.

is passed obliquely around the limb upward; these
terminate below the ankle in stout webbing.

When the plasters are applied, the limb is bandaged
in order to secure apposition of the plaster. This bandage, however, remains but a short time, and the next
day is removed, the skin being left bare except where
covered by the plaster. In this way, the circulation of
the limb is less interfered with than if a bandage were

worn continually, and eczema is less liable to occur. The plaster, however, needs to be renewed frequently, and rarely should be allowed to remain more than a fortnight or three weeks. The plaster can best be removed by washing the limb in alcohol or benzine. If there are excoriations, the plaster can be re-applied to

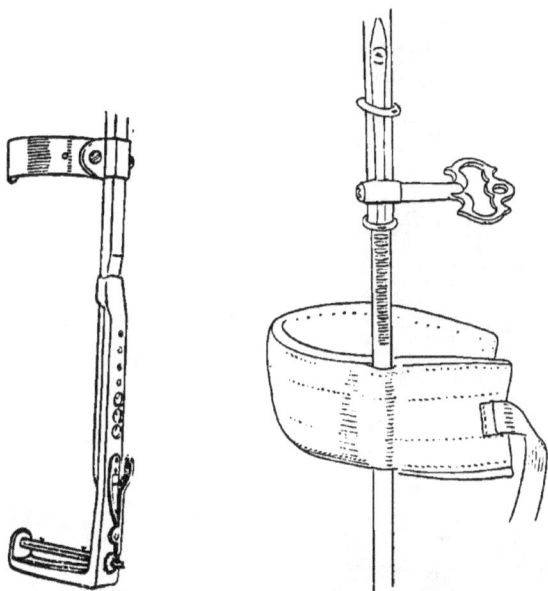

WINDLASS AND RATCHET APPLIANCES FOR EXTENSION.

[*From the Fiske Prize Fund Essay.*]

another portion of the skin, or the excoriated part can be protected by a cloth covered with zinc ointment.

Eczema is treated by powdering the skin with biniodide of bismuth. If, however, eczema becomes very annoying, the plaster-traction sometimes needs to be entirely removed, and a substitute for this, stocking-traction, is used. A stocking, with the foot cut off, is applied from just above the ankle to half-way up the thigh, and this stocking is cut longitudinally along the front. At the sides stout webbing is sewn, which serves to connect the stocking with the traction-appliance. Buckles or eyelets are sewn along the front of the stocking parallel to the cut part, and at a short distance (an inch) from the edge. The stocking is thus buckled or laced over the limb (cotton being applied next to the limb to prevent irritation), and a sufficient hold is thus secured to give a satisfactory pull upon the limb. Although as firm a hold cannot be secured in this way as by plaster, it is generally sufficient.

LONG TRACTION APPLIANCE.

The traction-splint consists of a pelvic band, a steel upright, and a traction-attachment. The pelvic band should be fitted to the pelvis in such a way as to make no pressure upon the bony prominences unprotected by muscles, particularly the anterior superior spines. The rod should be sufficiently strong to bear the weight of the patient with but little bending. The limb is secured to the rod by circular bands, furnished with straps, which pass behind or in front of the limb. A plate is furnished in front, which presses upon the upper portion of the thigh, and in this way prevents the splint from falling backward, and the pelvic band from striking on the anterior superior spines. The splint should be made and fitted accurately, and efficient traction should be applied.

Where traction-splints are used, perineal straps should be employed. These are made of stout webbing padded at the points of pressure with saddler's felting, and cov-

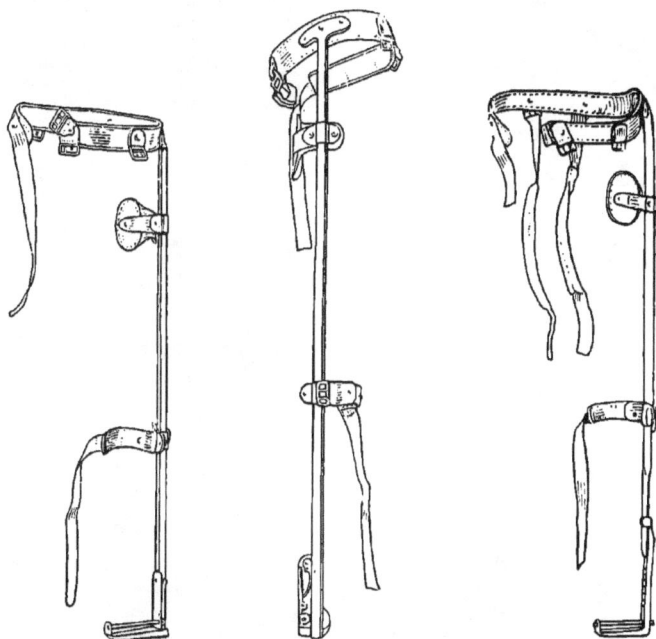

FORMS OF THE LONG TRACTION APPLIANCE.

[*From the Fiske Prize Fund Essay.*]

ered with cloth. Leather perineal straps, moulded on the perineum, are used in exceptional cases.

It is always the custom at the hospital to use the traction-appliance in connection with crutches and a raised shoe, as the patients are more thoroughly protected from a possible jar, or from the diminution of traction from the bending of the splint when weight is thrown upon it

in walking.   It is also the belief of the writers that
strong traction is essential during the earlier stage of
the disease.

After patients have passed the stage of acute inflammation, and when it is believed that a stage of conva-

CONVALESCENT SPLINT.

lescence has been reached, absolute fixation is not necessary.   Later still, a stage is reached where traction is
not needed to overcome the muscular force.   The traction-splint may therefore be discarded, and protection
alone furnished.   For this, crutches and the raised shoe
will suffice for adults.   With children, these are frequently laid aside through carelessness, although the hip

is not sufficiently strong to withstand without danger the jar of walking. Under these circumstances, an ischiatic crutch is of use. This consists of an apparatus similar to the traction-splint, except that it is inserted at the lower end into a socket fastened to the shoe. The rod is longer than the patient's limb, and when the weight of the body is thrown upon the shoe in walking, it falls on the perineum and not on the hip-joint. When the

WALKING WITH A CONVALESCENT HIP-SPLINT (Brackett).

foot leaves the ground, some pressure comes upon the hip, but this is slight, and occurs at a stage of the step where least protection is needed. This may be seen in the accompanying illustration. The ischiatic crutch can be made both with a hinge at the knee-joint, or without. The latter is less expensive than the former, but not as convenient. The value of the ischiatic crutch in the treatment of the convalescent stage of hip-disease is often overlooked. It is principally to careful management of patients in the stage of convalescence that good results in treating hip-disease are due; for in the convalescent stage, where no pain is present, children are careless, and they are exposed to jars from falls while at play. It is the prevention of relapses which renders a complete cure possible, and for this prevention protection is desirable until the epiphysis has become sufficiently strong to bear the jars incident to walking. For this, in growing children, at least two years are needed after the disappearance of all active symptoms.

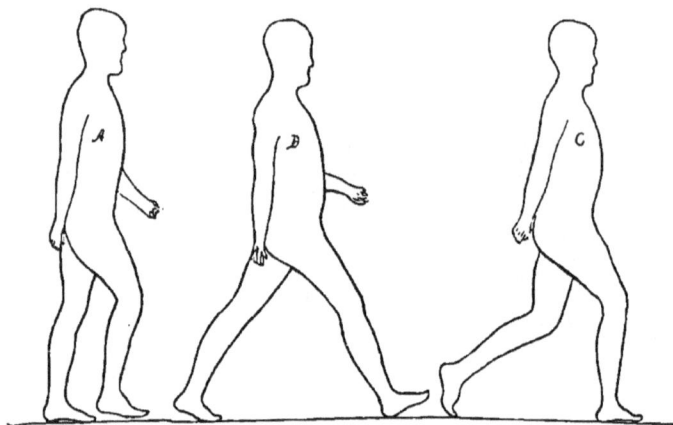

THE MECHANISM OF WALKING (Brackett).

NIGHT-CRIES. — Night-cries are frequently met with in the early stage of the disease, and should be regarded as indicative of acute inflammation in the joint. Under these circumstances, the patient should be carefully guarded from jar, by providing that the limb shall be thoroughly fixed and protected. As a rule in cases of this sort, when placed under recumbent treatment, with traction, night-cries disappear in a few days. This, however, is not always true, and cases occur where night-cries persist for weeks. In such cases, the formation of an abscess may be expected, and can usually be detected within a few months after the subsidence of night-cries. After the active stage has passed away, this symptom does not recur. The frequency of night-cries in the exceptional cases where they persist, may possibly be diminished by phenacetine, salicylate of soda, or chlorodyne.

DEFORMITIES. — The deformities which occur in hip-disease are treated according to the stage at which they occur, and according to their severity. Flexion due to reflex muscular spasm is the most common. This, as well as abduction and adduction, is corrected by fixation in bed, and traction applied to the leg in the line of the deformity (vide page 164). As muscular spasm subsides under rest and traction, the raised support to the limb is lowered, and in a short time the deformity will be found corrected. After subsidence of the acuter symptoms, ambu-

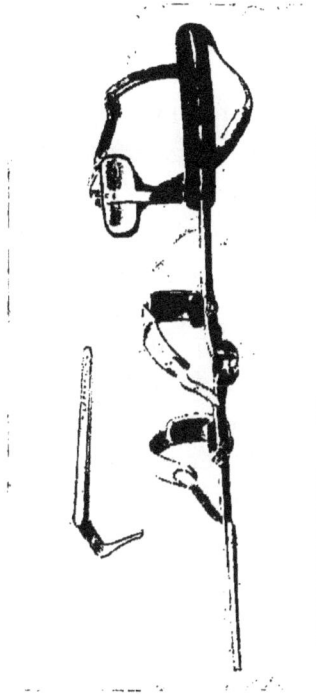

CONVALESCENT SPLINT JOINTED AT KNEE.

latory treatment is resumed. This method of treating deformities is efficacious in a later stage of hip-disease, but in such cases a longer time is needed for correction. The traction-splint will be found of great help as furnishing a thorough means of traction. The correcting force in adduction is furnished by the perineal straps attached to the pelvic band.

In cases which will not yield under traction in bed,

forcible correction, or osteotomy, is needed. The former has been done in a number of cases. It is not attended with danger, provided ordinary care is used. After correction, the limb is fixed by a plaster-of-Paris spica bandage, and the patient kept in bed on a frame for several weeks. After this, locomotion, crutches, a high shoe, and a traction-splint are used. In older cases, where fibrous or osseous anchylosis has taken place, osteotomy is necessary. This has been performed twelve times at the hospital. The results have been satisfactory in all cases. The method of operating is as follows : Without a preliminary incision, the osteotome is pushed directly through the skin, the width of the blade being in the line of the long axis of the thigh. As soon as the bone is reached, the osteotome is turned so that the blade is at right angles to the axis of the femur, and driven three-quarters through the bone. The limb is then broken by manual force. There is no bleeding of importance, as there are no large vessels in the vicinity. A simple aseptic dressing is sufficient.

After the operation, the limb may be fixed by a plaster-of-Paris spica bandage, but fixation on a bed-frame with weight and pulley has been found to be more convenient.

ABSCESSES. — During the years of 1884 to 1892, inclusive, there were 717 cases of hip-disease, and in 153 of these the abscesses were incised. The treatment adopted at the hospital has been almost invariably that of incision. Aspiration and expectant treatment have been carried out in but a few cases. The abscesses are incised in such a way as to give complete drainage. Two or more incisions are made if complete drainage of the sac demands it. The wall of the abscess is curetted, and the cavity washed out thoroughly, and wiped with iodoform gauze. Iodoform wicking is inserted, and the wound dressed aseptically.

The course of the abscess is found to be modified by the use of the traction-splint, which, as it furnishes pressure in the perineum, in the iliac region, and on the front of the thigh by the pressure-plate, effectually prevents burrowing of the abscess in these directions. If the abscess extends, it must enlarge in the front or the outer side of the thigh at a point favorable for incision. When the abscess has been thoroughly drained, recurrences are exceptional. Sinuses, due to the persistency of the carious process, frequently remain for a long time in a certain number of cases.

c. OPERATIVE TREATMENT. — Incision, excision, and amputation have all been employed at the hospital. Incision has been, however, performed but a few times. In five instances, it was done to relieve extreme pain and to prevent sleeplessness at night, on the supposition that there was fluid in the joint. This, in three cases, was not found to be the case, and no relief was obtained. In one case alleviation was noticed. In one instance, incision was made and a curette used, but no benefit was obtained. In another case, incision was made to remove a loose sequestrum in the advanced stage of hip-disease. This was found to give relief, and the patient recovered. Erasion of the joint has been done a few times, but without marked benefit.

Early excision of the hip-joint has been done but a few times. This is owing to the fact that facilities are provided in the institution for the continuous treatment of cases for a long time. Under these circumstances, conservative treatment is usually followed by favorable results, and it is believed that, where conservative treatment is possible for a long time, the results are better than when excision is employed. In the few instances in which early excision has been done, the results have been excellent. In some cases, excision has been delayed too long, and a cure has not been accomplished.

The incision used has usually been that of Langenbeck; in a few, Hueter's incision was used. Complete excision (*i. e.*, removal of the bone below the small tro-

WARD WAGON.

chanter), with curetting of the acetabulum, has been the rule. The wound-cavity is stuffed with iodoform-gauze, and an aseptic dressing applied. The patient has been treated, after the operation, upon a fixation-frame, with light traction. In some instances, the Thomas splint has been used instead of traction; but traction-treatment has been preferred.

There have been thirty-eight cases of excision performed at the hospital between 1878 and 1893. Of these, five died within a short time after the operation. The deaths may be considered as directly due to the operation. There were also thirteen deaths from general causes after the operation. One died from nephritis; eleven from general tuberculosis; in one the cause was not known. Two of these underwent amputation two months after excision; recovered from the amputation, but died a year later. Of the remaining twenty, six have not been heard from. Fourteen are known to have

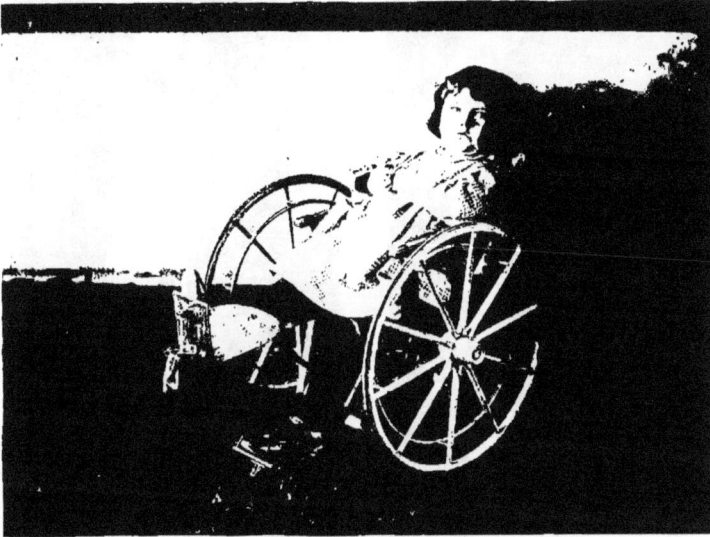

WARD CHAIR FOR HIP-DISEASE.

recovered. Of these, one, after undergoing amputation at the hip-joint, recovered, although extensive pelvic caries was present. At present, ten years after the amputation, he is in excellent health.

The known duration of the disease before excision, in the fourteen reported cases, varied from five months to five and a half years, the average being about two years. The head of the femur was found to be diseased in every case, and the neck in six cases. The acetabulum was diseased in twelve of the cases, and was perforated in four cases. The time which has elapsed since operation varied from eleven months to ten years. The results at the present time, as far as they can be tabulated, are, general condition, good in nine cases, fair in

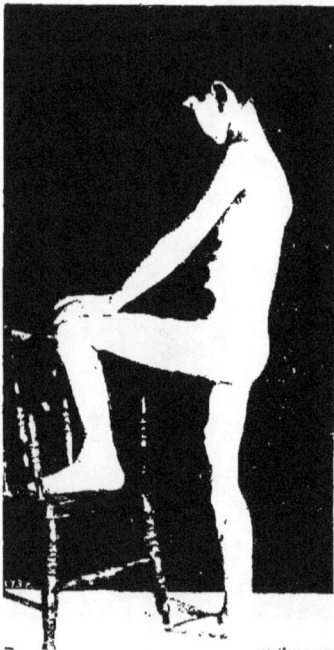

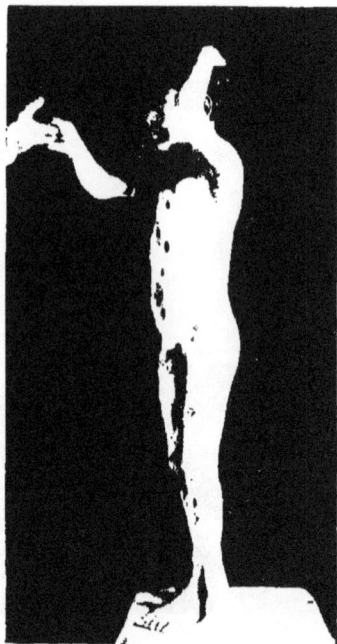

(a) A case of hip-disease under ambulatory treatment. *Result good.* Motion to right angle in flexion.

three, and very poor in one. Six are left with discharging sinuses.

The splints used were the ischiatic crutches applied as a protection to prevent relapse.

AMPUTATION. — Amputation following excision has been done in three cases. The patients recovered from the operation. Two, however, are known to have died, a few months later, from amyloid degeneration of the liver and kidneys. One recovered completely.

The method of controlling hemorrhage was by means of an elastic band. A circular incision was made with a long lateral incision, and the bone removed subperiosteally. Reformation of bone took place in the stump of the case which survived.

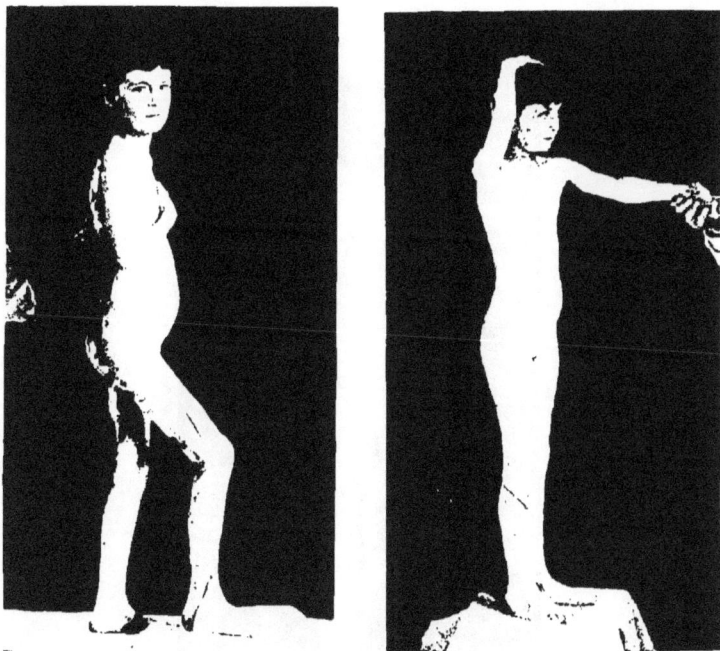

(b) A case of hip-disease under ambulatory treatment. *Result fair.* Motion to 45° in flexion.

DOUBLE HIP-DISEASE. — In a few instances double hip-disease has been treated. The double Thomas-splint has been found serviceable. Weight-and-pulley traction can be applied while the patient is in bed, and the traction-attachment can be aided by perineal countertraction in the more acute stages.

CONCLUSIONS. — It is impossible to define accurately how long the various methods of treatment are pursued

in the various stages of the disease. Treatment is begun
as soon as a diagnosis is established. Where the symptoms are acute, and there is much distortion, bed-treatment is advised. Where there is muscular spasm,
traction is employed. Where there is sensitiveness, fixa-

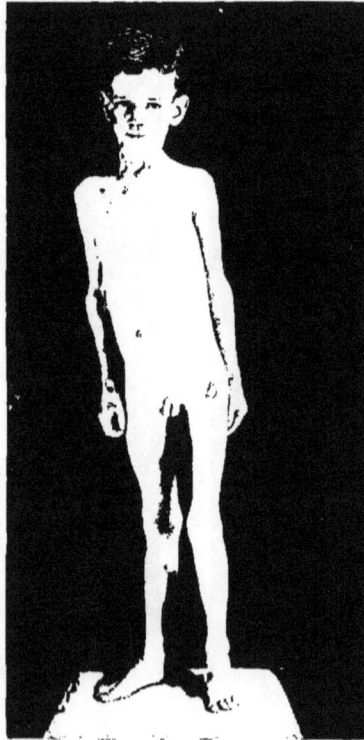

(c) A case of hip-disease under ambulatory treatment. *Result bad.* No motion
at hip, discharging sinuses, shortening and wasting of limb.

tion is used, and protection is regarded as necessary as
long as any treatment is indicated. The length of time
protection is continued depends upon the age of the
child, its weight, and rate of growth.

In the first stage, treatment is vigilant; in the acute

stage it is rigorous; and in the convalescent stage it is careful. In cases where treatment has been undertaken in an early stage, and thoroughly carried out, ultimate recovery, with a useful limb, is confidently expected;

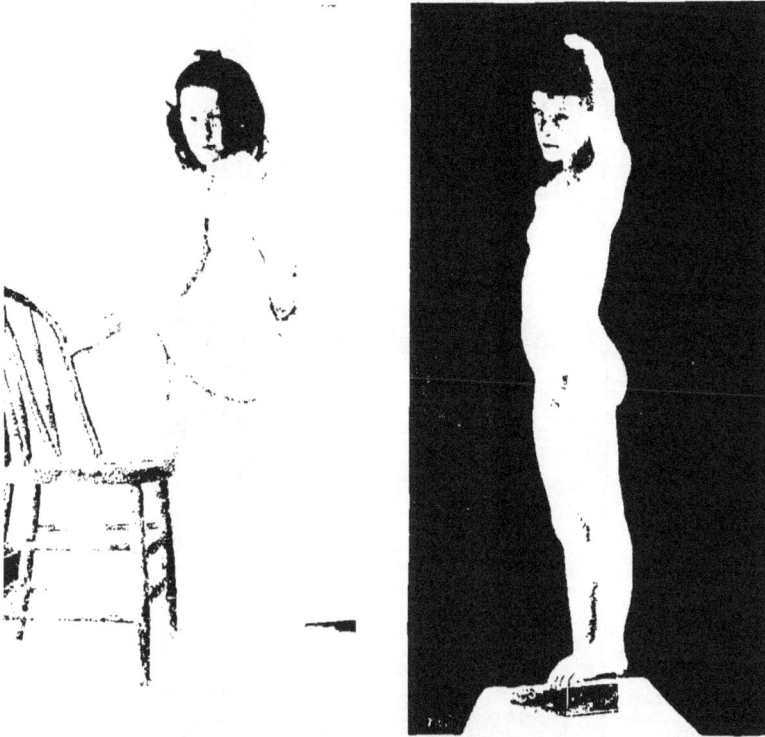

(d) K. C. A *good* result after late excision of the limb. Disease treated conservatively for two years. Now motion good in all directions, flexion, abduction, and external rotation at a right angle. Sinuses healed. Three inches shortening. All evidences of active disease have disappeared. Five years since operation. Walks without apparatus.

but the excellence of the cure depends partly on the thoroughness of the treatment, and partly upon the extent of the tubercular process. In some cases, perfect cure, with re-establishment of motion, has been obtained.

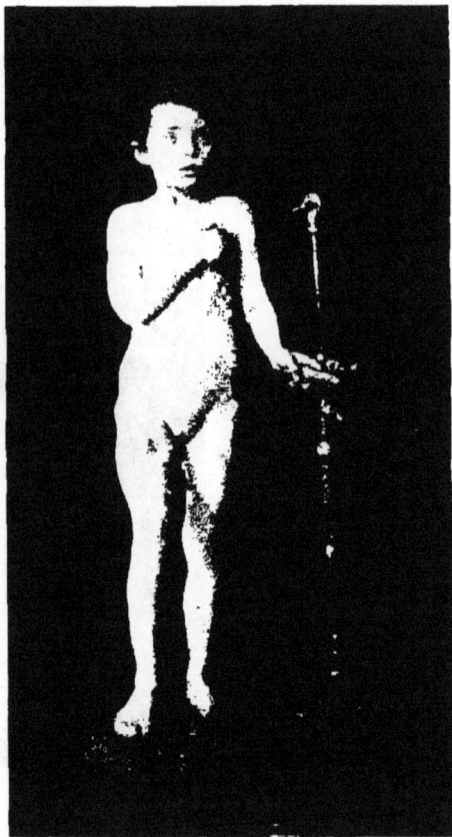

(e) M. O. A *fair* result after late excision of the hip. Disease treated conservatively for three years. A few degrees of motion in flexion, other motions painful.
Sinuses discharging. Two and a half inches shortening. General condition poor.
Walks without apparatus, but limps badly. Four years since operation; condition
has not improved in last year.

[*From the Fiske Prize Fund Essay.*]

(*f*) J. D. A *poor* result after excision of the hip. Disease treated conservatively nine months. (A very painful acute case, with rapid abscess formation.) Now no motion at hip. Sinuses healed, but hip painful. Walks with long traction-splint. Three years since operation.

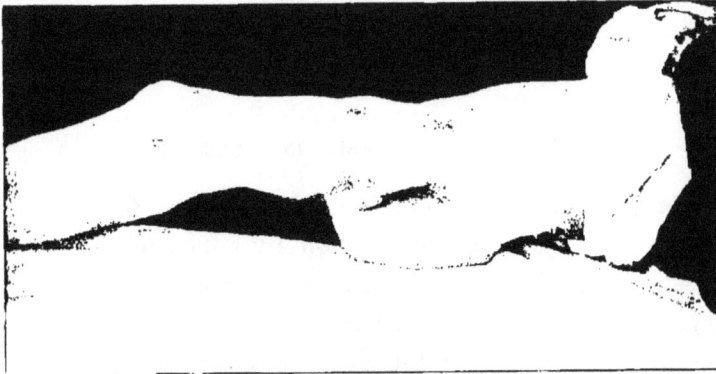

(*g*) M. S. A *bad* result after excision of the hip. Disease treated conservatively for one and a half years. Now hip abducted through an arc of nearly 45°. External rotation of nearly 90°, flexion to 60°. Four inches shortening. Anchylosis complete. Deformity great, and cannot walk without crutch. Sinuses healed. Nine years after operation.

[*From the Fiske Prize Fund Essay.*]

E. H. B.
R. W. L.

DISEASE OF THE KNEE.

DISEASE of the knee-joint is treated mainly by the methods of protection and fixation, with the addition, in cases of particular sensitiveness, of traction by means of adhesive plaster. The methods employed depend upon the character and the severity of the case, and may be grouped into —

    1. The Acute Stage of Tumor Albus;
    2. The Convalescent Stage of Tumor Albus;
    3. Cases with Deformity;
    4. Operative Cases;
        a. Arthrectomy;
        b. Excision.

The apparatus employed for carrying out the treatment consist of —

    The Thomas Knee-Splint;
    The Plaster-of-Paris Bandage;
    The High Sole and Crutches; and
    Traction by Adhesive Plaster.

1. TREATMENT OF CASES IN ACUTE STAGE.

Cases in the acute condition are treated by fixation by plaster-of-Paris bandage, traction by adhesive plaster in cases of especial sensitiveness and tendency to deformity, with rest in bed when necessary. Protection is obtained by means of the Thomas splint and high sole and crutches.

2. Treatment of Convalescent Cases.

In the convalescent stage, the treatment consists of protection by means of the Thomas splint and high sole and crutches, fixation of the leg by leather bands secured to the splint, and, in the later stages, the application of the caliper-splint, which is used as a walking appliance.

3. Treatment of Cases of Deformity.

Very few cases of angular deformity are treated as out-patients for this condition itself; it is preferred to send them into the hospital for the correction of the deformity, which is accomplished either by traction or operation. In the cases in which the deformity has lasted but for a short time, and is due to muscular spasm, the leg is brought into the extended position by means of gradual traction, carried out as nearly as possible in the line of the deformity.

4. Treatment by Operation.

Operation is performed in those cases in which the deformity has existed for a long time, and is not due to muscular spasm, but to adhesions and contractions about the joint as the result of disease. The operation consists either of forcible correction, or, in cases where there is anchylosis, of an exsection of a wedge-shaped piece of bone. The forcible correction is performed by a genu-clast, in which the force is exerted upon the tibia in a direction to correct the subluxation and the flexion at the same time. The instrument is shown on page 190, and has the advantages of an adjustable plate, " so that by moving the arch *a*, upon which the power is applied, nearer to or farther away from the knee, it can be used

upon a child or an adult. The application of the power
in front, instead of behind, is more convenient and more
easily managed. The screw *b*, working in the arch *a*,

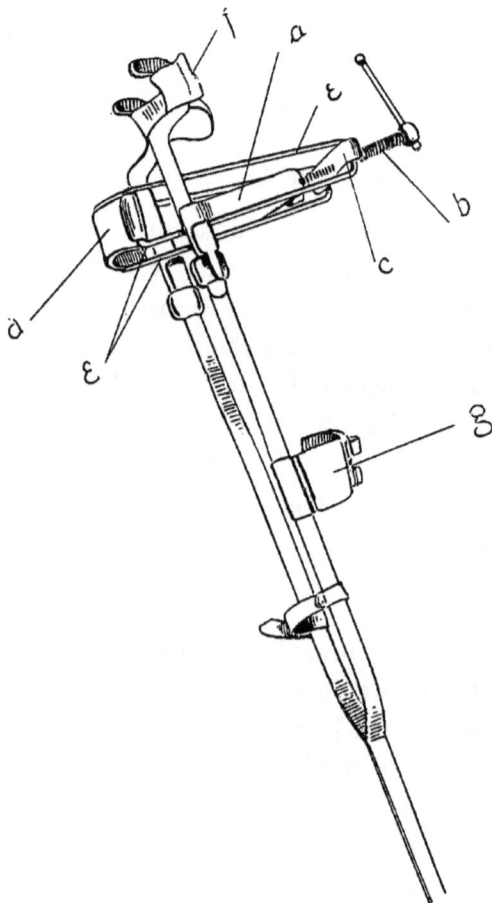

GOLDTHWAIT'S GENUCLAST.

[*Boston Med. and Surg. Journal.*]

raises the crossbar *c*, to which the posterior band *d* is
attached, by means of the steel loops, *e*. The counter-
pressure comes upon the end of the femur by means of
the leather pad *f*, and to a less extent upon the strap *g*."
In using the instrument, "the plate *c* is forced forward,

carrying the head of the tibia with it, and the counter-pressure coming on the end of the femur by the straps *a*. After as much has been gained as is possible in this way, the leg is straightened, the strap *d* still being the ful-crum, the head of the tibia is drawn forward into its normal position." The leg is then placed in a plaster-of-Paris bandage, with or without extension, as may seem necessary, and kept in it for three or four weeks. A protection splint is then applied, and worn until there is no evidence of tendency to a return of the deformity. When the hamstrings are found so contracted as to oppose the correction, tenotomy is performed.

In cases of long anhylosis from extensive disease, and when the disease is too advanced to offer hope of suc-cess by conservative methods, or in acute cases of exten-sive disease, when the care necessary for a long period of time cannot be given by the parents, operation is ad-vised. The conditions which indicate operation are the involvement of bone, the degree of deformity, and the activity of the process. In cases of extensive disease which offer no hope of cure with motion in the joint, a more complete operation is done, and the diseased por-tions of the bones are removed with the saw, the ends of the bones brought together, and the leg placed in a plaster-of-Paris bandage. When the disease is found to be less diffuse, and presents the opportunity of thorough removal without resorting to an excision, the operation of arthrectomy is done.

THE APPLICATION OF THE THOMAS SPLINT.

The Thomas splint is practically the original Thomas splint. It is shown in "The Thomas Knee Splint" and "Applied" (vide page 192). It consists of an oval ring

which encircles the upper part of the thigh, which is joined
to two iron wire uprights which extend downwards along
the outer and inner sides of the leg, to two or three inches
below the sole of the foot, and are here joined together
by an oval ring which serves as a foot piece, and a bar

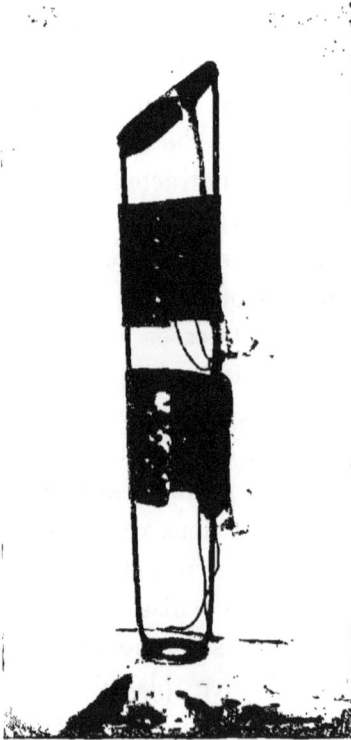

THE THOMAS KNEE SPLINT.   THOMAS KNEE SPLINT APPLIED.

which is used for extension when needed. The inner
upright is joined to the oval hip ring at an angle of 55°,
which is padded with felt and leather so as to reduce the
angle to 45°. This may be used as a crutch, the patient
resting on the tuberosity of the ischium. The leg is
secured to the splint by means of leathers which encircle

both the thigh and the calf, being attached either to both
or to one upright, and are made to lace in front of the
leg. In some cases, the method of Thomas is used,
which consists of stretching a band of leather, three or
four inches in width, behind the leg and securing it to

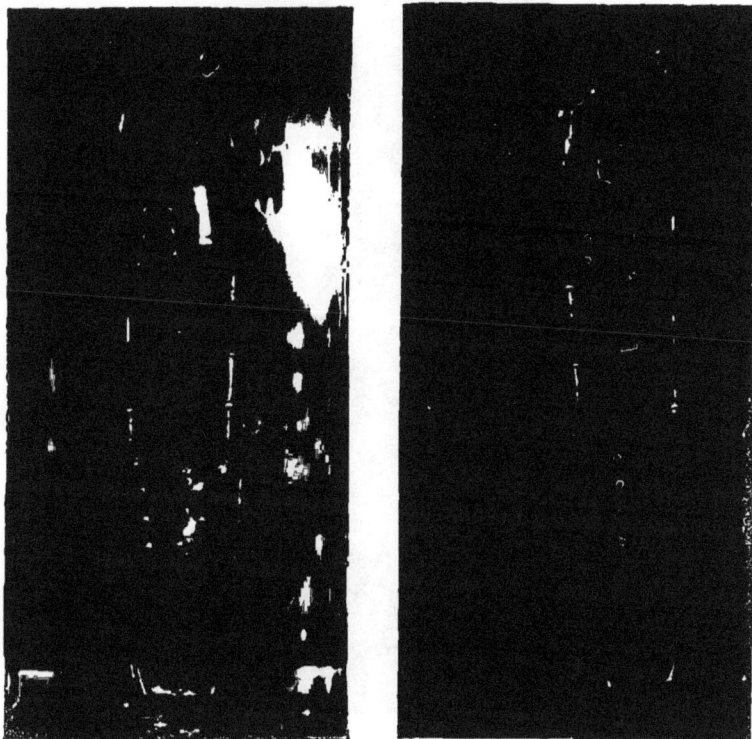

THE THOMAS KNEE SPLINT, WITH LEATHER LACINGS.

both the uprights, so that the upper extremity of the tibia
rests upon the upper border of the leather, and a second
band of leather is in front of the thigh, with the lower
border just above the patella, and the leg is then further
secured by means of a bandage round the ankle. The
splint is shown above.

13

In cases requiring less protection, the inner half of the hip ring is cut away, and the ends joined to the uprights by a curved rod of iron, so as to leave a space between the upper end of the inner upright and the tuberosity of the

THOMAS SPLINT WITH INNER PART OF RING CUT AWAY, FITTED
AS A CONVALESCENT SPLINT.

ischium; and from the two extremities of the cut ring is slung a perineal band, on which the patient rests in the same manner as in a hip-splint. The oval ring on the lower end of the apparatus is removed and the ends of the upright inserted in a socket in the heel of the shoe. The length of the splint is so adjusted that when the patient bears the weight on the perineal band, the heel of the foot does not touch the sole of the shoe, but the ball of the foot is allowed to come to the ground. By tension on the perineal band, the amount of weight borne by the foot may be regulated. The splint is shown on page 194.

THE APPLICATION OF THE THOMAS SPLINT AND PLASTER-OF-PARIS BANDAGE.

The plaster-of-Paris bandage is applied with the leg in a position which it assumes without force, and is carried from the lower third of the leg to the upper third of the thigh. The Thomas splint is applied in the same manner as when leathers are used to secure fixation, except that both bands of leather are made to encircle the leg, as the fixation to the uprights is not so necessary as when the leathers alone are used.

If less fixation is required, the plaster may be split along its upper surface, and eyelets sewed parallel to the margin. This cast is then laced together, and forms a secure fixation-splint, but may be taken off for bathing the limb and for rest. The same treatment may be carried out by a leather splint moulded over a cast of the leg, and has the advantage of greater durability.

THE APPLICATION OF ADHESIVE PLASTER.

When traction is desired, the adhesive plaster is applied in the same manner as in hip disease, by means of narrow strips of rubber adhesive plaster placed along the outer and inner side of the leg, with cross-encircling straps, but all are carried only to the level of the knee. The traction is secured by passing the strap from one side around the bar at the bottom of the splint, and buckling it to a strap on the other side, and the amount of traction is secured by the tension on this strap.

The high sole is used in all cases, and is high enough to raise the splint clear when crutches are used, and, when the splint is used as a crutch, to make both of equal length.

THE CONVALESCENT SPLINT.

The convalescent splint is intended to give fixation to the knee, but freedom to the leg in walking, and at the same time the patient is allowed to bear a certain portion of the weight on the diseased leg. The upper part of the splint may be of the same pattern as that used in the acute stage, but the lower ends of the uprights are bent at a right angle, and fitted to a socket in the heel of the shoe, and are of such a length that, when the weight is thrown on the splint, the heel does not touch the ground, but the patient is allowed to use the ball of the foot. If more freedom is desired, the modified hip ring, with the perineal band, may be used, which allows more motion of the leg in the splint. The leathers are applied in the same manner as in the earlier periods of treatment.

TREATMENT OF CASES OF SIMPLE OR TRAUMATIC SYNOVITIS OF THE KNEE.

Cases of this character are not frequent, but are usually seen after injury, and present symptoms of pain and swelling from effusion, and usually some heat and tenderness. In these cases, if they appear to be benign in character, the knee joint is fixed by a posterior ham-splint, and compressed by means of bandages of either cotton, gauze, or rubber. When necessary, cool applications are used. This treatment is continued as long as inflammatory symptoms and swelling are present; after which, the joint is given rest, either by a protection splint and the use of crutches, or else by rest in bed until all danger of secondary trouble is over.

THE TREATMENT OF TUBERCULAR SYNOVITIS.

No distinction is made in the treatment of this condition from that of the acute stage of tumor albus.

CASES OF INFANTS.

A condition in infants is occasionally met with presenting an acute inflammatory trouble, attended by bony enlargement, decided tenderness and heat, with pain, redness, loss of mobility, and usually permanent flexion. A large number of these cases come to abscess formation, but the course is benign, and a recovery usually occurs in the course of five or six months. As these cases are too young for ordinary splint treatment, they are usually supplied with a Cabot wire frame, or the rectangular gas pipe frame, and the leg is secured at rest at whatever angle seems to be best adapted to the condition of the knee at the time. In this way, patients are carried about in the same way as hip or spine cases, and this is continued as long as the joint requires local fixation, as shown by the pain, tenderness, and loss of mobility. After this, the infants are kept in arms without a splint, and are not allowed to crawl about until all symptoms have disappeared, and the knee has resumed its normal shape and its usual function.

E. G. B.

DISEASE of the ankle-joint is treated at the hospital

(*a*) By conservative methods,

(*b*) By operation.

Conservative treatment is applied in the out-patient department, in all cases of average severity, even in those of long duration. All early cases, even if severe, are treated in the same way. This treatment is persisted in until it is manifest that it is powerless to control the disease. As a rule, the results of conservative treatment are satisfactory.

Operative treatment is at times undertaken in very acute and severe cases with abscess formation, even in fairly early stages; more often, however, it is reserved for cases with much thickening and sinus formation, for excessively painful cases, and for cases where the destructive inflammation of the joint has persisted in spite of carefully applied conservative measures. Abscesses are opened as they occur. Early excision is not practised, but the operation is kept as a resource for cases where conservative treatment has failed.

The number of cases of ankle-joint disease in the out-patient department is not large, as may be seen from the figures taken from the annual reports.

1888	16 cases.
1889	10 "
1890	9 "
1891	12 "
1892	12 "
	59

During this time fourteen cases were admitted to the wards, many of them being included in the above list.

(*a*) The success of conservative treatment requires the two conditions, — (1) *fixation*, (2) *protection*.

(1) *Fixation* is obtained most often by a light plaster-of-Paris bandage which includes the leg from below the knee to the toes. Where it is possible, the foot should be fixed at a right angle; but when malposition of the foot, due to muscular spasm, is present, and the foot is held either too much extended or too much flexed by the irritated muscles, the plaster bandage is applied to the foot in the deformed position, without any attempt at forcible correction. The quieting effect of fixation is such that the muscular spasm relaxes and the foot is easily fixed by the bandage at a right angle to the leg, after one or two plasters have been applied.

Fixation is continued until the heat and joint irritability have subsided, and have been absent for some time. Even at this period of the disease, it is not considered safe to bear weight upon the diseased joint, and protection is continued for a still longer period.

The plaster-of-Paris bandage is occasionally split down the median line in front, and is furnished with lacings (as in the case of a removable plaster jacket) so that it can be removed daily, much to the comfort of the patients.

An appliance which is often used in place of the plaster-of-Paris bandage is shown in the figure on page 200, and fixes the joint more definitely. It is, moreover, more comfortable in being lighter and cooler. It is sometimes applied merely as a fixation appliance in the class of cases in which one might use plaster-of-Paris. It consists of a steel sole plate made to fit the sole of the foot, and two uprights which run from this plate, one on each side of the leg. These are connected at the top by a posterior steel calf band. A leather lacing holds the foot snugly to the plate, and an upper lacing fastens the uprights to the leg. It is applied over the stocking, and is worn inside of the shoe. There is no joint at the ankle, and the splint is only worn in connection with a protection appliance. Fixation of the ankle-joint, in some instances, is furnished by a moulded leather splint, by a right-angled splint made of tin, or by a posterior wire-splint bent to a right angle.

FIXATION ANKLE-SHOE.

(2) *Protection.* — Inasmuch as fixation is obviously only a part of the treatment of ankle-joint disease, it becomes necessary to secure the second and equally important part of the treatment, — protection of the ankle-joint from bearing the body-weight in walking. Except in the case of infants in arms, the use of the Thomas knee-splint, or some similar protection appliance, in *addition to* the apparatus for fixation, forms part of the routine treatment. The Thomas splint is applied just as in knee-joint disease, and independently of the fixation appliance, whether plaster or steel. Its object is merely to keep the diseased ankle from bearing any weight.

In the adult, a high shoe on the well foot and crutches are generally quite enough to prevent putting the diseased ankle to the ground; but in the case of children, some further protection is necessary on account of their thoughtlessness.

Protection is continued until after the indications for fixation have passed away, and is only given up after heat, swelling, muscular irritability, and all other signs of disease in the joint have ceased for some time. Then the splint is gradually discontinued, while the patient is under observation.

Disease of the metatarsus is treated in practically the same way as if the tarsus were involved, except in cases where the ankle-joint is in no way restricted in motion. If in these there appears to be no need of fixation, the Thomas knee-splint is alone applied to protect the ankle from the jar of walking. If such cases progress poorly, fixation is at once added to the treatment. The results of this treatment of metatarsal disease have been almost uniformly satisfactory, even where pus has formed and broken through the skin.

(*b.*) Operative measures. — *Abscesses* are opened and curetted, and then packed with iodoform gauze. If dead bone is present, it is removed.

Erasion of the ankle-joint is not practicable, because of the complicated surface over which the synovial membrane spreads out.

Excision. — The results of the excisions of the ankle done at the hospital were studied in 1889,[1] and analyzed, showing, perhaps as well as anything can, the experience of the hospital in this regard. 54 cases had been operated upon, between 1869 and 1889, of which it was pos-

[1] Charles L. Scudder, Trans. American Orthopedic Assoc., Vol II. p. 53.

sible to speak of the results in 18 cases, which were
not recent or unhealed cases. Of these 18 cases, 6 had
died, 5 of tubercular meningitis; but in the 12 cases that
recovered, the results were all reported as good. From
an analysis in detail of the cases reported, which is published with the paper referred to, it is easy to see that
late excision may yield, and does yield, the most admirable results. In some of the cases, the result was nearly
perfect; and in two instances the difference between the
arc of motion in the diseased ankle and in the well one
was only eight degrees. The shortening of the leg in
most of the cases was slight, and in many of them there
was none at all. In most cases the sinuses were healed.
These cases will serve to show the value of late excision,
practised only in the severest cases, and where conservative measures had failed, having been first tried in nearly
all of these cases.

From 1889 to 1893, inclusive, there have been 11
cases of ankle-joint disease operated upon in the hospital.
During this time, 50 to 60 new cases have applied at
the out-patient department, in addition to the cases
under treatment from former years, which shows how
small a proportion of cases have required operative
treatment. Of the 11 operative cases, there were 9
excisions and 2 curetting operations. In 4 cases, secondary operations were necessary; and in 3 of these, several
operations were necessary to eradicate the disease. One
patient died of tubercular meningitis three weeks after
operation.

Two cases excised in 1890 were doing well when last
heard from; the sinuses had healed, and there was good
motion. Two cases excised in 1892, now (in 1894) have
good ankles, with sinuses healed and with good motion.

Some patients could not be found, and the others (operated since 1892) are too recent to be of value from a statistical point of view.

No one method of excision is practised to the exclusion of the others, but the Kocher operation is much liked. This consists in cutting the peroneal tendons and outer ligaments of the joint, and dislocating the foot inward; this gives an admirable view of all parts of the joint. In many cases, two lateral incisions have been used, while again, the incision has been made wherever the site of the disease seemed to warrant it.

The operation, whatever method is pursued, is done bloodlessly as far as possible, so that the diseased tissues may be more readily identified.

After operation, iodoform wicks are loosely packed in the wounds, and the foot is retained in position by a light plaster-of-Paris bandage applied over the dressing, or an accurately shaped posterior wire-splint is sterilized and applied over a few layers of gauze, and the major part of the dressing is put on outside of this supporting splint.

In any event, immobility of the ankle-joint is secured for at least some weeks. When the child is well enough to go about, the posterior wire-splint, or some similar one, is applied to immobilize the ankle-joint, and a Thomas knee-splint is used to keep the foot from touching the ground.

R. W. L.

DISEASES OF THE ELBOW.

IN disease of the elbow-joint the conservative method of treatment is followed. Fixation is obtained either by a tin internal angular splint, or by a snug plaster-of-Paris bandage extending from the finger-tips to the shoulder. If the elbow is held rigidly in a position of deformity, the fixation bandage is applied in that position.

The position for fixation, chosen in all cases, especially as anchylosis is not unlikely to occur, is at a right angle.

A certain class of cases which might be spoken of as suspicious often apply for treatment. These are cases where, some weeks after a fracture or a sprain of the elbow, pain, heat, and limitation of motion occur in connection with more or less swelling. Inasmuch as some of these cases have been observed to eventuate in tuberculous disease of the joint, the plan is pursued of considering them serious, and of treating them accordingly.

Excision is done as a last resort. If abscesses form, they are opened and curetted. Free incisions are made in much swollen elbows, and these measures are in general preferred to the formal excisions in the case of young children.

The results of conservative treatment have been as good as could be expected. The majority of cases have recovered without suppuration, some with movable joints. One case died of generalized tuberculosis, and one or two feeble children have died of exhaustion.

R. W. L.

DISEASES OF THE WRIST.

DISEASE of the wrist has presented itself chiefly as a tuberculous synovitis and ostitis. A chronic teno-synovitis has been seen in a few cases, presumably of a tuberculous nature. This latter has yielded more readily to treatment than have the affections of the joint proper.

Cases of wrist joint disease have been treated by rest and fixation on a plain wooden splint, with counter-irritation over the joint; or they have been dressed with millboard splints over sheet-wadding, and subjected to uniform pressure, after Gamgee's method. This exercises a steady compression upon the diseased tissue which the writer believes assists in its absorption. This form of fixation is therefore preferable to the simple palmar wooden splint. Plaster-of-Paris affords excellent fixation, but is heavy and less suited to the use of children, and leather splints, although good, are expensive. A sling, to support the arm and keep it quiet, is a necessary part of the treatment.

The majority of cases under conservative treatment do well. Permanent thickening almost always results, but the wrist is rarely stiff. Fixation is continued until the disease is quiescent.

If suppuration occurs while the case is under treatment by fixation, a partial or complete excision is done, as seems indicated; but unsatisfactory results have, as a rule, followed extensive excisions of the wrist, and operation is therefore deferred until conservative treatment seems to be useless.

R. W. L.

CLUB-FOOT.

TALIPES EQUINO-VARUS.

SINCE the establishment of the hospital, all the methods at present recommended for the treatment of club feet have been tried, from the Scarpa's shoes to wedge shaped excision. The practice, however, at present is well defined, and the deformity has become one which is treated with precision and uniform success. Treatment must necessarily vary according to the age of the patient and the condition of the deformity. In the accompanying report, only the congenital variety of equino-varus deformity is considered; the paralytic is referred to elsewhere. The basis of this report is a hospital experience of upward of 400 cases of congenital club-foot.

INFANTILE CASES.

Treatment is begun as soon as the nutrition of the child gives reason to believe that continuous treatment can be carried out. The practice is to treat these cases in the simplest way. This consists of the application of a plaster-of-Paris bandage to the foot held in the corrected position, the bandage being applied to the foot and leg, and above the knee, attention being paid to the correction of the varus deformity first. When this is sufficiently overcome, the tendo Achilles is cut. The foot is then held in an overcorrected position by a plaster-of-Paris bandage, which is worn for ten days or a fortnight. After this, a varus shoe is worn, applied to the corrected foot by means of a silicate bandage. After this, if the correction is satisfactory, and no contracture is to be felt either in the plantar fascia, tendo Achilles, or tibiales muscles, a retentive appliance is worn where fixation is accomplished by means of straps. This is worn for a year or more, and is so constructed that it can be applied within the shoe, and allows motion of the foot in every direction except that of inversion, or dropping of the front of the foot. Treatment by means of the retentive appliance should be continued as long as there is any tendency of the foot to assume a faulty position, — ordinarily until the child is two or three years old. If the treatment has not been thorough, and any of the fibres of the plantar fascia, or of the shortened tendons, have been

VARUS SHOE.

left undivided, or if the subsequent overcorrection has not been thoroughly done, a partial relapse occurs, and a second operation is necessary. This should be done before the child is three years of age; but a second operation is never necessary if the earlier treatment has been thorough. Infantile cases can be corrected without

tenotomy, but in hospital practice this method of treatment is tedious.

OLDER CASES.

In children between three and six, more resistance is
met than in infants. The practice at the hospital has

been immediate rectification after tenotomy. The foot
is forced, under an anæsthetic, into an overcorrected
position. The hand, or a Thomas or lever wrench, is
used to complete the correction. If, however, obstinate
resistance is encountered, Phelps's open incision is used,

and all resistant tissues are divided. In the early experience at the hospital, sloughs were occasionally met with under the retention plaster bandage; but at present, with better attention to detail, these rarely occur. Bandages are made of unsized crinoline gauze into which dry dental plaster-of-Paris is incorporated. The bandages are three yards in length, and one and one half inches in width.

Before applying the plaster bandage, the foot and leg are wound with sheet wadding bandages; over this, the ordinary wet plaster-of-Paris bandage is applied, the foot and leg being held by the hand of the surgeon. Two or three bandages are quickly applied, and, before they have stiffened, the foot is grasped firmly and forced into as nearly a corrected position as possible. For the purpose of correction, the leg is held by one hand and the foot by the other; the foot is grasped with the sole resting upon the palm of the hand, the great toe being pressed upon by the inside of the thumb; the fingers reach around the foot, and rest upon the prominent astragalus. If the foot and toes are well protected with cotton, there need be no danger of sloughing.

The correction should be chiefly in the direction of eversion of the foot before any attempt is made to correct the equinus deformity. In order to prevent the plaster from being soiled when hardened, it can be covered by shellac, asphalt paint, or paraffine.

Bandages, in older as well as in younger children, always reach above the knee, which is slightly bent, otherwise the bandage will twist around the limb, and eversion will not be completely corrected.

It should be borne in mind that, after a club-foot has been thoroughly corrected, it is necessary that the foot

14

be retained in a corrected or overcorrected position for some time, in order that the contracted tissues shall not re-contract, and the misshapen facets of the astragalus, scaphoid, the os calcis, and cuboid shall have time to adjust themselves to the normal condition, and alter into the normal shape. It is, therefore, necessary that a retention apparatus be worn for some time. Various forms of retention appliance have been used. The one

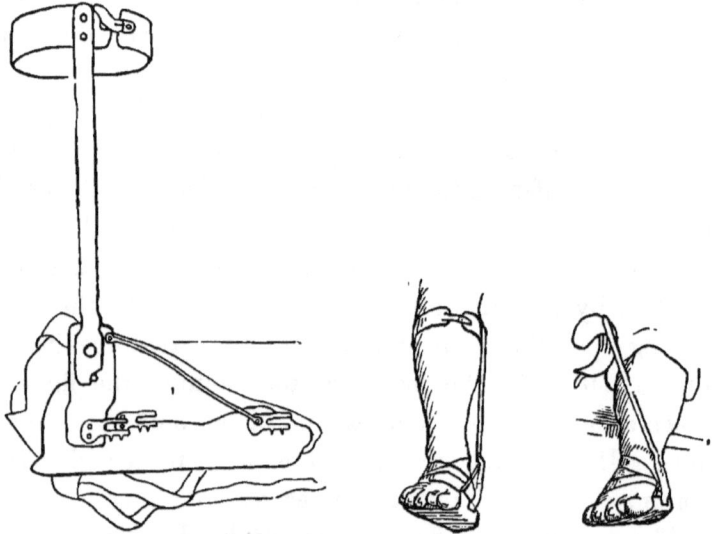

RETENTION APPLIANCE, UNAPPLIED AND APPLIED.
[*From Bradford and Lovett.* — *Orthopedic Surgery.*]

in general use is similar to that used in young children, and is depicted in the accompanying illustration. The foot can be secured by straps; but if the mothers are negligent, or the foot resistant after removal of the plaster, it is found better to retain the foot in the splint by means of a silicate bandage wound about the foot and ankle. If the plate is protected by means of felt, there is no danger of chafing.

In cases with a tendency to twist, the apparatus should

go above the knee, and take support around the hips. A boot can be worn over the apparatus; an ordinary laced boot being used, opened down to the toe, which, if laced over the apparatus, furnishes an additional means for fixation. It will be borne in mind that it is essential that the heel of the child presses well down upon the sole plate when the latter is at right angles with the axis of the leg; if that is not secured, relapse is apt to take place. This retention apparatus should be worn night and day for some time, usually a year, sometimes longer, and should be gradually laid aside. Treatment can be supplemented by massage and electricity, if desired, obviating in some cases the necessity of a long continuance of the apparatus in many cases; but as the apparatus permits motion in many directions, muscular atrophy is not to be dreaded. The appliance should be continued until the patient walks, without apparatus, on the flat of the foot, and without toeing in. Relapse may occur in cases not properly attended to, either from imperfect correction or a lack of thoroughness in the application of the retention appliance.

In children under five, manual or wrench correction is usually sufficient to correct deformity; but in children older than this, and in some of the more resistant cases, Phelps's method of open incision is of advantage. It was the former practice at the hospital to rely upon forcible mechanical correction; but of late years the advantages of open incision have become so manifest in resistant cases that it is employed in the severer cases. The disadvantage of open incision is the greater delay in healing than if subcutaneous division is made, but this is offset by the greater thoroughness of division in the severer cases. It has been found that the open incision should be thorough, dividing the skin and all the resistant parts, especially the ligaments, between the scaphoid and astragalus. No trouble has occurred from dividing the artery in the sole of the foot, the cross incision through the sole of the foot being made sufficiently far forward so as not to endanger the tibial artery before its division to the smaller branches. Three cases have suppurated from a fault in the asepsis. In these, healing was delayed, the foot swelled, suppuration took place sufficient to demand counter-incision on the dorsum of the foot. By careful antiseptic treatment, soaking the foot daily in antiseptic solution, the suppuration was corrected, and a cure resulted. This accident, however, could be avoided; and, in cases where it has occurred, it was in all probability due to the imperfect cleansing of the foot before operation. In a few cases in children older than ten, even after open incision, complete correction was not possible. A certain amount of resistance was encountered at the articulation of the cuboid with the os calcis. In these cases, wedge shaped osteotomy of the neck of the os calcis, and in some instances also of the neck of the astragalus, was necessary.

Excision of the astragalus was done in five cases, in the belief that this would obviate the necessity of after treatment. In one of them, however, a relapse occurred, which was only corrected by an osteotomy of the os calcis. In the others, the results were not as satisfactory as those which follow open incision with or without osteotomy of the neck of the astragalus and os calcis.

In one case, a simple osteotomy of the neck of the astragalus, followed by forcible correction, was sufficient to make a complete cure of a relapsed case. In another, however, this failed, and an open incision, with osteotomy of the neck of the os calcis, effected a cure.

In two cases, wedge shaped exsection of the tarsus was performed with excellent ultimate results. They were cases met in the early practice of the hospital; subsequent experience demonstrated the superiority of methods involving less mutilation; *i.e.*, osteotomy of the neck of the astragalus or of the os calcis.

Excision of the cuboid alone has not been performed.

Adult or adolescent cases are not included in this report. They have been treated, in the experience of the surgeons connected with the hospital, according to the same principles as above described, with the exception that osteotomy of the neck of the astragalus and of the os calcis is usually required in resistant adult cases.

Indications for omitting treatment involves the question of the treatment of cases during what may be called the convalescent stage. These are largely matters of judgment, and cannot be defined by precise rules. This stage varies from a year to two or three years.

Night appliances are worn for two months following the operation; after that, they can usually be discontinued. As a rule, the larger the patient, the shorter the time protection is needed; for at every step, the weight of the patient acts as a corrective force, provided the foot is placed well in position. It may be understood that unless the foot is thoroughly corrected, the stage of protection is necessarily prolonged; because if there remains a vice of formation, protection is necessary until the malformation is outgrown or corrected by surgical means. If the deformity is thoroughly corrected, it does not take long for pressure to bring the fasciæ into their normal state, — provided this is not prevented by any contracted ligament or tendon.

As a rule, in infantile cases after complete correction, a retention apparatus should be worn night and day for six months. Walking appliances should be worn for a year or two after this. In children of five the same is true, except that walking appliances are usually needed for a longer period than a year.

In older children as well as in adolescents, walking appliances are needed for a year after correction. Where open incision or osteotomy has been employed, the need of a walking appliance for more than six months is exceptional.

E. H. B.

H. W. C.

INFANTILE PARALYSIS.

THE general treatment of this affection varies according to the severity of the case. In the less severe forms electricity and massage are used alone; in the severe cases, where there is much atrophy or deformity, these measures are used in connection as subsidiary to orthopedic or surgical treatment. The ordinary electrical treatment is Faradism to the affected muscles, which is given twice a week for not more than three minutes at a sitting. This treatment is in severe cases often continued for months, with excellent results. Of late, massage, given by an experienced masseuse, has been regularly used in a large proportion of cases, both alone and followed by Faradism, and in suitable cases forced movements are also employed. No internal medication is given.

What may be termed the surgical or orthopedic treatment of infantile paralysis is directed to the prevention of deformity, and the correction of such deformity, if it occurs, by the employment of such apparatus as enables the paralyzed limbs to be used in the most available way. Surgical treatment is necessarily confined largely to the deformities of the leg and foot, as paralysis of the upper extremities is comparatively infrequent, and is not attended by disability severe enough to attract the attention that it would in the leg. The surgical treatment is usually undertaken after the stage of convalescence is past, when the patients have reached a condition where but little is to be expected from electricity or massage. In short, where it seems as if a continued use of the limb would result in deformity, or where deformity is already present, surgical treatment is considered necessary.

Where deformity is already present, two means are used for its correction, — tenotomy and mechanical stretching. In cases where tenotomy is required, it is of course most usual to find the deformity of talipes equinus along with flexion of the knee, and possibly of the hip. In these cases the tendo Achilles, or hamstring tendons, and the muscles below the anterior superior spine of the ilium, are the ones to be divided. In general, open incision is preferred to subcutaneous tenotomy, except in the case of the tendo Achilles. After tenotomy and forcible correction, the limb is straightened and fixed in a plaster-of-Paris bandage, which secures the limb in a corrected position.

Mechanical stretching, however, in children, is often efficient, and does away with the necessity for operative interference, as contractions yield easily, especially in the earlier stages. Again, simply bandaging a limb, or applying a splint, will often overcome deformity. The time needed for mechanical correction varies with the extent of the deformity. In the lighter cases three or four weeks only are necessary, while in severe cases weeks or months may be needed. With tenotomy, the time is shortened. After tenotomy in infantile paralysis there is not so much tendency to recurrence as after tenotomy for congenital affections, such as talipes equino varus.

In any case of infantile paralysis of the leg, whether deformity has been present or not, the problem is to so control the leg that it shall be used as nearly as possible normally. A support is needed in the case of the lower extremity to prevent the knee from dropping forward.

Apparatus for the purpose of merely keeping the leg straight may be very simple, or it may be complicated.

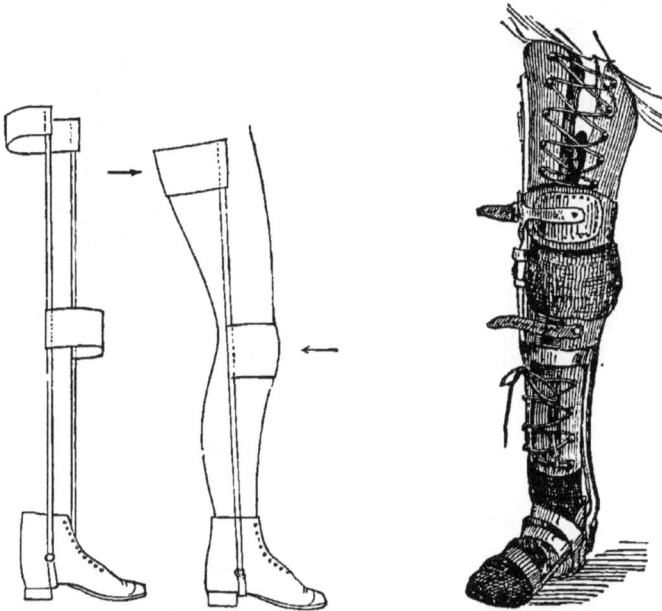

APPARATUS FOR INFANTILE PARALYSIS.
FOR KEEPING THE LEG STRAIGHT.   APPLIED WITH CORSET.
[*From Bradford and Lovett. — Orthopedic Surgery.*]

The illustrations show the simplest form of apparatus. It has no joint at the knee, but one can easily be furnished without great expense. Where there is deformity of the foot, as is the case in most patients, some supporting appliance embodying the same principle is applied to the leg in connection with the shoe for equino varus, or some similar supporting appliance for the foot.

In the illustration shown, the simple drop catch is used at the knee. Where the muscles of the trunk are affected by the paralysis as well as the lower extremities, any appliance for use must be reinforced by the use of a stiffened leather or paper corset. In some of the se-

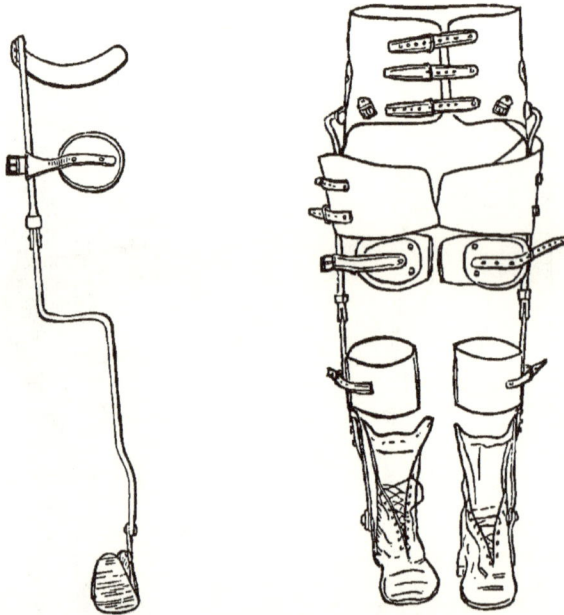

APPARATUS FOR INFANTILE PARALYSIS.
WITH TALIPES.   BURRELL'S APPLIANCE.
[*From Bradford and Lovett. — Orthopedic Surgery.*]

verest cases the use of such an appliance as this has enabled patients who were reduced to the condition of quadrupeds to walk about with the use of crutches. This, of course, necessitates a certain amount of training, and consumes a great deal of time. After the patients have learned to adjust themselves to the appliances, the crutches can sometimes be laid aside, and in some instances the corset may finally be dispensed with.

Any apparatus which tends to induce a more normal use of the limbs or the back tends to improve the nourishment and muscular condition of the parts so used. A leg, for instance, which is atrophied and contracted may be seen to increase in size and improve in circulation when once such an appliance has been put on as will necessitate the use of the leg in the proper plane and in the normal motions. The fitting of such an apparatus, and such improvement, naturally consume much time and require patience. It is rarely the case in infantile paralysis that all the groups of muscles of the lower extremity are paralyzed, although they may have that appearance. An electrical examination, however, will demonstrate the existence of certain operative normal muscles among them, and upon the development of these weakened and improperly used muscles one depends largely for the usefulness of the leg.

Severe cases, where both legs and possibly the lumbar muscles as well are paralyzed, are fortunately rare. The apparatus shown on page 218 is intended for the use of those patients where it is necessary to supply even a hip joint. A stiff leather pelvic band may be prolonged upwards into the corset if necessary. In the lighter cases, improvement may be expected in the development of the strength of the affected muscles if the deformity which causes the weakened muscles is corrected and prevented, provided that treatment is begun before the muscles have undergone too great fibrous degeneration. The treatment of deformities of the foot alone following infantile paralysis is similar to the treatment of the same deformity of congenital origin, except that, as has been mentioned, relapse is less likely. Tenotomy of the contracted muscles has in no instance been followed by any ill effects, and the use of the equino varus shoe is often efficient, in the milder cases effecting a cure.

A detailed account of the methods of treatment in deformity of the foot alone will not be entered upon here.

E. H. B.
W. N. B.

HERNIA.

THE hospital records of the last ten years show a list of three hundred and ninety-seven cases of hernia, principally of the inguinal variety.

The treatment of these cases has been Mechanical and Operative.

THE MECHANICAL TREATMENT.

This is employed with out-patients only. The child is fitted with an appliance which prevents the recurrence of the hernia without causing atrophy of tissues by pressure, and which can be worn without discomfort. A majority of these patients can be furnished with such an apparatus. The forms used are: —

1. The "Pye" worsted truss;

2. The ordinary spring truss, with single or double pad;

3. Modifications of 2;

4. Atypical appliances, consisting of flat pads of some rigid substance. Wood, pasteboard, leather, hard rubber, aluminum, and several other substances have been used. These pads are oval, round, or triangular, as is best adapted to the special case. They are held in place by belts, straps, swathes, or bandages, as is most efficient.

Patients are treated by this method when the hernia can be controlled by the apparatus; when too young to be admitted to the "In" service; when the parents refuse to allow operative treatment; when the patient can receive at home the care necessary to make this method of treatment successful.

Each new patient is carefully examined to ascertain the exact anatomical condition of the region affected. The form of apparatus is then selected which is best adapted to control the hernia and fitted to the patient. The parent is next definitely instructed how to carry out the treatment; how the apparatus should be cared for; what it is intended to do; how it should be applied; how and when it should be worn. The child is brought at short intervals to the hospital until it is seen that the truss is efficient, and that the parent has learned how to use it correctly. The patient is then required to return every six to eight weeks for inspection. If at any time the apparatus is not efficient, or anything is unsatisfactory to the parent, the child is brought at once to the surgeon.

This treatment is continued for twelve to eighteen months. When this time has elapsed without a recurrence, the truss is gradually omitted, at first while in bed, then during the day while quiet. Finally its use is wholly discontinued.

It is important to determine the real value of this mechanical treatment. Some surgeons claim that it is effective and cures without the danger of an operation, while others believe that in most cases it is a waste of time, except in so far as it is palliative.

This question cannot be answered positively from the data furnished by the records, on account of the great difficulty in ascertaining the permanent results, and also because these children belong to parents whose indifference, indolence, ignorance, or poverty is a great obstacle to success. The occasional cases of children whose parents can and will devote the necessary time and attention to them give much more satisfactory results. Notwithstanding this, it is certain that at least six per cent of the patients treated as above described have been permanently cured.

This statement is based on the results of 285 cases treated at the hospital during the past ten years.

17 were cured.

229 " relieved.

15 " unrelieved.

24 treatment not carried out.

The patient is considered cured when the following record has been obtained: The patient has been treated without recurrence for twelve to eighteen months; after gradually discontinuing the truss the child has been under observation for six to eight months without recurrence; that the child was known to be well at the time the data for this report were collected, (August, 1893). Six per cent is a minimum, and if it can be obtained under such unfavorable conditions as have existed with most of these patients, it is probable that a larger proportion could be cured if the children were so situated as to receive the necessary care and attention.

The duration of this treatment in the successful cases is usually a period of months, in some cases, years.

In regard to apparatus it is difficult to specify, except in a general way, what form is to be used. What is satisfactory for one child is inefficient for another. The appliances which have given the greatest satisfaction are —

The "Pye" truss, especially for infants;

The simple pelvic spring truss and pad, sometimes with, sometimes without a perineal strap;

The same, with an especially centred aluminum flat pad devised by Dr. E. G. Brackett;

A "pad" truss, which consists of a firm, flat, triangular pad held in position over the internal inguinal ring and canal by straps radiating from it in four directions, viz., right and left, forming an horizontal band around the pelvis; upward over the shoulder of the affected side, and down the back to the horizontal pelvic band; downward across the perineum, and upward to the horizontal band. This has been efficient. It is not expensive, and can be worn without discomfort. It is especially satisfactory for infants and very young children.

THE OPERATIVE TREATMENT.

This method, if successful, permanently cures the patient. It may, even when this result is not attained, so improve the patient's condition as to make truss treatment possible, and thus relieve where mechanical treatment alone is impossible. Operative treatment also seems to give superior opportunities to attain what natural processes have failed to accomplish, even when aided by artificial support. The surgeon can actually see what defects exist; he can more intelligently remove them; he can effect a cure much more rapidly than by the mechanical treatment.

At first, in this hospital as elsewhere, operative treatment resulted in more failures than successes; but methods and technique have gradually been so perfected that now, to the individual patient, a permanent cure can be promised in a majority of cases. The permanent cure is rapidly obtained, and the patient made independent of surgical care and apparatus.

Those patients are advised to submit to operation with whom the hernia cannot be satisfactorily controlled by apparatus; with irreducible herniae; in many cases of undescended testis; when the care and attention necessary to make mechanical treatment successful is absent; when a child's social condition is such that its adult life will probably be a laborious or exposed one; when eighteen to twenty months of correct mechanical treatment has failed to effect a cure.

The methods used have been those of Heaton; Macewen; Macewen with various modifications, including Bishop's; Bassini; McBurney; and "Atypical" opererations.

Of these, the Heaton has not been satisfactory, — one success in nine operations. The herniae usually recur a few days after the operation. It has not been performed at the hospital since 1887.

The "Macewen" operation as originally described, or with various minor modifications, has been more frequently done than any other, and has given satisfactory results. It is at present the "routine" operation. The wounds are sutured without drainage, and usually heal under the first dressing, which is removed on the twelfth day.

The Bassini has been used in only in a few cases, too recently to furnish data as to a permanent cure. There have been no recurrences.

The McBurney operations have not been especially satisfactory. The healing of the wound is longer than by the other methods; also, the absorption of the cica-

15

trix has been rapid and marked, in one case leaving a thin wide scar.

There has been no result among the Atypical operations so strikingly successful as to entitle them to especial mention.

All operations are done in a thorough aseptic manner. Carefully sterilized silk is used for the sac and inguinal rings; silk or catgut for the superficial sutures. To protect the wound from urine, the dressing is covered with rubber or protective, or its outer layers saturated with Tr. Benzoin Comp., or antiseptic collodion. At times the wound itself has been sealed with the benzoin or collodion.

As stated above, the successes obtained warrant one in promising a permanent cure in a majority of cases. This is shown by the following data: Of 51 operations there were 43 immediate cures. These operations have been done during the past six years. At the time the permanent results were investigated, August, 1893, accurate information could be obtained from only 25. Of this group, 19 were permanently cured. These, tabulated, were: —

	Cases.	Cured.	Recurrences.	Deaths.
Macewen	12	8	3	1
" (modified)	4	3	1	0
" (Bishop's modification)	4	4	0	0
McBurney	2	1	1	0
Atypical	3	3	0	0
Total	25	19	5	1

This gives 76% of permanent cures, which can probably be increased by the present improved technique, and increased knowledge and skill resulting from the work of the past ten years. Such results warrant the surgeon in making the above promise to his patient.

The routine after treatment is rest in bed for from four to six weeks. After the wound is healed, a light pad and bandage is worn for several weeks to support the affected area till organization of the new-formed tissues is complete. Until the wound is healed, the patient is kept at rest by a rectangular bed-frame. The pain resulting from the operation is seldom severe, and usually ceases after twenty-four hours. The average time in the hospital ward has been four weeks, but many patients have been kept in bed six.

It has been observed that recurrence is more liable to occur in patients whose tissues are relaxed, thin, or atonic; where large herniae have caused marked dilatation of the inguinal rings and canal, or when early cessation of bed treatment has exposed to tension recently healed tissues before organization is complete.

When well, usually at the end of six weeks, the child is transferred to the Out-service, where it is frequently seen. Any exercise or work liable to cause sudden intra-abdominal pressure, as jumping, climbing, or lifting weights, is forbidden for the next eight months. If at the end of this time there is no recurrence, the child is discharged as well, but is ordered to return at once if anything abnormal is noticed. Should the hernia recur, a second operation is done. In one case a third operation finally cured the patient.

H. W. C.

LATERAL CURVATURE.

SPECIAL provision has been made for the treatment of cases of lateral curvature, so that patients under the age of fifteen are received in the Out-Patient Department, but none are received into the House over the age of twelve, and but few are given house treatment for this condition. The clinic is divided into three departments, each of which is under the direction of some one person who works under the supervision of the surgeon in charge. These three departments are: 1. Recording; 2. Gymnastic Treatment; and 3. Apparatus Treatment. Each case as it presents itself has its history carefully taken. The patient is then measured, the strength tested and recorded, a comparison made with the tables of average growth and development, and then a series of exercises, and whatever apparatus treatment may seem adapted to the case are prescribed.

In the history there is taken note of any antecedent family and personal data that may have a bearing upon the condition, and includes the history of the health and the growth during early childhood, the date of the appearance of the curve, its rapidity of increase, the general condition and strength of the patient, and any circumstance which may have occurred during the early life which would have a bearing upon the development of the deformity. The child is also examined for any asymmetrical development, as a difference in the length or development of the legs, flat feet, &c. For convenience in taking these data and measurements a printed card is used.

Name		Age	Date
Address			Height.............
			Weight.............
Age noticed			Circumferences
			Neck...............
			Chest normal, R ....
Previous growth			" " L ....
			" full.......
Rate of increase			" empty....
			Abdomen........
			Pelvis ....
Subsequent growth			Biceps, R........
			" L........
			Breadth
Heredity			Shoulders........
			Hips ....
Sc. f.	Meas.	Diphth.	Lengths
			Arm, R...........
			" L...........
Thoracic disease			Leg, R...........
			" L...........
Paralysis			Depths
			Chest, R........
			" L........
Cause given by parents			Abdomen........
			Strength
Habits of position			Back...........
			Abdomen........
			Side, R.........
Occupation			" L.........
			Deltoid, R......
General Health			" L......
			Scapular, R.....
			" L.....
Diet			Lat. Dor., R....
			" L....
			Pectoral, R.....
Early rickets			" L.....
			Grip, R.........
			" L.........
Hip	Knee	Foot	Flat foot, R
			" L

RECORD.

The record begins with measurements of the height, weight, muscular strength, and general measurements of the body, all of which are compared with the table of averages, and the proportion noted. The character and degree of the deformity is recorded by marking the outline of the curve on the diagram of a symmetrical figure, showing the changes in the shape of the trunk, which include changes in the level of the shoulders and the scapulae, and of the depth of the waist-arm angle. The amount of deviation of the spinous processes, from the line of the seventh cervical to the fold of the but-

tock, is measured, both in the standing and lying position, and the amount of rotation, as shown by the
deformity of the thorax, is recorded by the cross tracing
of the thorax at the desired level, and this also is taken
in the standing and lying position. In addition, each

THREAD FRAME FOR RECORDING DEVI-
ATIONS OF THE SPINE.

case is photographed in a
standing position through
a thread frame, which
is divided off into inch
squares, by which the
measurement of any de-
viation from the normal
can be taken from a pho-
tograph, except that of
the rotation of the thorax,
and also the difference in
the curve of the spine, as
shown in standing and
lying positions, to indicate
the amount of flexibility.

No arbitrary division of
the cases is made so far
as the treatment is con-
cerned, but for conven-
ience in description they
may be grouped into three
classes, which differ main-
ly in degree: 1. Flexible cases of slight deformity in
which no bony change has occurred; 2. Cases in which
there has occurred a certain amount of bony change,
but which are still flexible; and, 3. Older and resistant
cases in which there has been considerable bony change,
and reduction is not possible.

The means employed in the treatment of the different cases are: Gymnastics, Mechanical and Forcible Correction, Apparatus.

First Class of Cases.

Cases coming under the first of these classes are treated by gymnastics alone, or, if necessary, with the addition of a light brace, which is only postural in object, — that is, a brace which is applied to prevent the patient from assuming those mal-positions which are found to be common in this class of patients. The gymnastics given are in part symmetrical movements carried out in the self-corrected position, and asymmetrical movements which are selected for the individual cases. Apparatus which is used for these cases is not of any particular pattern, but usually is some light form of steel appliance, sometimes fitted to the corset, and which has for its object simply the prevention of faulty attitudes.

Second Class of Cases.

Cases included in the second class are treated by exercises and the application of retentive apparatus. The gymnastics used are usually those of special selection, having mainly for their object the increase of flexibility, such as those which increase the backward flexibility of the spine against the direction of the deformity, and increase the rotation of the thorax in the direction of the normal plane. These gymnastics are carried out once or twice daily, according to the circumstances of the patient and as often as it may be necessary they report to the hospital for observation, and for the necessary forcible and mechanical correction. The apparatus used in

these cases is usually some form of retention corset, —
either the removable plaster-of-Paris, leather, or paper

Third Class of Cases.

Cases which are included in this class are those in
which the curve has become more or less fixed, and in
which there is less hope of improving the condition,
and frequently all that can be attempted is the prevention of an increase of the deformity.   If the curve is
rigid, and if, from the circumstances of the patient, there
seems to be but little possibility of improving the curve
by the strict attention to daily exercise, a retention corset is given in case the deformity shows a tendency to
increase.   If more time can be given, the cases are
then given a strict exercise treatment, to be carried out
both at the hospital and at home, and which has for its
object purely the increase of flexibility, and, in addition
to this, a retentive corset which is applied during the
mechanical correction or over-correction, and which is
made over a cast taken with the patient in this position,
and is made to exert as much corrective force as is com-
fortably borne.   In patients where the gymnastic treatment cannot be carried out at home, and where the
case is of such a character and of such age that during
an added growth an improvement may take place, the
application of permanent plaster jackets, frequently
changed and put on during over-correction, is attempted,
and in a certain number of these in which the treatment
has been faithfully carried out, a decided improvement
has been shown.

GYMNASTIC TREATMENT.

With the exception of the few cases wearing permanent plaster-of-Paris jackets, all patients have more or less gymnastic treatment, which is prescribed at the clinic and carried out at the hospital and at home. Patients living within access come to the hospital on three days during each week, and are given individual attention by some person trained in this work; and so much of the exercise as can be done either alone or with the help of the parents, is carried out daily at the home.

The gymnastic treatment as employed in this clinic has a twofold object: first, the improvement of the general muscular condition, which is given for the general development and training; and, second, that which has for its object the increase of flexibility, and should properly be considered under the head of forcible correction. The first, or the lighter forms of gymnastics, are usually symmetrical, and are given either as body movements, or body movements with apparatus work, — such as chest weights, etc.

Those exercises for increasing flexibility, and which are purely gymnastic in character, consist of forced body movements only; while those exercises employed through the means of apparatus are considered under the head of mechanical correction.

FORCIBLE CORRECTION.

This part of the treatment is done at the hospital only, and for this the patients present themselves at regular intervals. The exercises of forcible correction consist of the application of pressure with the object of increasing the flexibility of the spine, and correcting the rotation. These are all passive in character, and are used either by long continued force, and applied and regulated by screw force, or as forcible exercise given by the assistant, with or without the aid of mechanical means.

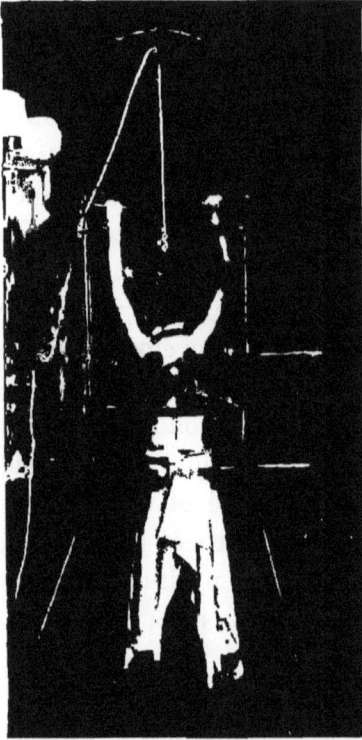

STANDING PRESSURE CORRECTIVE MACHINE.

The mechanical correction is carried on by means of the pressure correction machines, of which there are two kinds, designed for pressure in the standing and in the recumbent position. The apparatus for the former consists of a large upright frame, in which the patient stands and is partially suspended by a head sling and by the hands. By means of clamps the hips are fastened so that no motion in this direction is possible, and shoulder pressure is made by adjustable plates. The rotating pressure force is applied by means of plates which are attached to long screws, and may be placed on any part of the patient's trunk, and pressure made in any direction. By means of the screw, the amount of the pressure can be accurately gauged. The patients are allowed to stand in this frame until they show signs of fatigue, and are given as much pressure as can be borne without discomfort, and while under treatment are made to breathe deeply. As a rule, from ten to twenty minutes is the usual time of treatment.

The recumbent apparatus has for its object the application of the same corrective force, and its principle is the same, except that the superincumbent weight is removed during the exercise, and suspension is not necessary. It consists of the hammock frame, such as is used

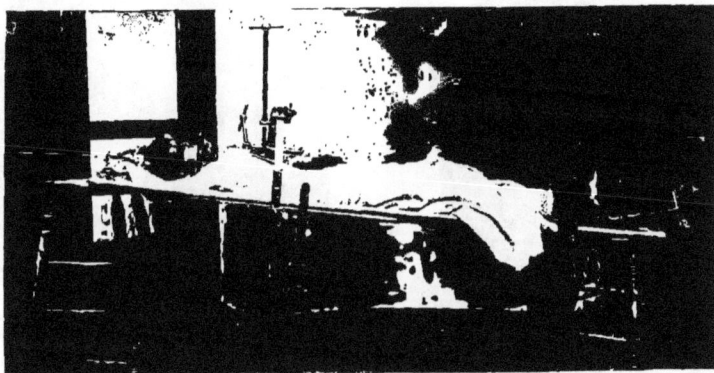

RECUMBENT PRESSURE CORRECTIVE MACHINE.

for plaster jackets, fitted with pressure pads, as is the upright apparatus. Its advantage is that fatigue is less, and the pressure is borne for a much longer time.

APPARATUS.

No description will be made of the lighter form of steel apparatus, as no particular pattern is used, but each case is furnished with whatever form of appliance seems best adapted to prevent the patient assuming mal-positions. The form of apparatus used for the severe cases is that of stiff corsets, either of plaster-of-Paris,

of leather, or paper, and made over a mould which has
been corrected, to give the desired amount of pressure.
In taking the plaster cast for this mould, no attempt is
made to correct the rotation of the thorax; but care is
taken to have the shoulders and hips in the same planes.
The cast from this is then changed so that a jacket made
upon it will make pressure on the prominent parts, and

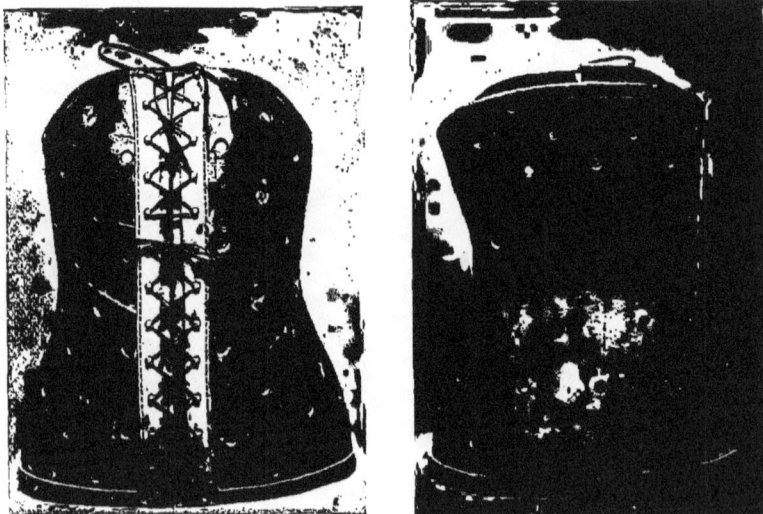

PAPER JACKET.

leave room for the expansion of other parts. The choice
of the kind of corset depends upon the circumstances of
the patient. The plaster-of-Paris removable jacket and
the leather jacket, however, do not exert so thorough a
rotating force as the paper jacket, which is made of sufficient thickness and stiffness to prevent yielding, and it
also possesses the additional advantage of not softening
under the moisture of the body.

Paper jackets are made from a corrected cast of the
patient, and consist of from five to six layers of matrix

paper pasted together in strips about two inches in width, which are placed over the cast in different directions, and between each two layers of paper is pasted a strip of linen. When dry, these form a firm support, and have the advantage of cleanliness, inasmuch as they can be washed without injury. They are always worn throughout the day, except during the gymnastic exercises, and may be worn at night.

E. G. B.

RICKETS.

THE cases of rickets which present themselves for treatment in the Out-Department may be grouped in three classes: (1) cases of general rickets; (2) rickets with deformity; (3) rachitic spines.

FIRST CLASS: CASES OF GENERAL RICKETS.

The first class, those cases presenting symptoms of general rickets, are usually found in young children or in infants, in which general symptoms are most prominent; that is, there is either weakness, retarded development, enlarged epiphyses, a soft condition of bone, or a bad general nutrition. The condition is usually found in those cases which have not yet walked, or walk but little.

The children are usually brought for retarded development rather than for local conditions, such as deformity of bone. The Italian and colored population furnish a large majority of these cases, and among these races the bony distortions seem to be most pronounced. As these cases at the time that they are brought under observation are generally backward in walking, the bone distortion, if such exists, becomes of much less importance for the time being, and in many cases can be disregarded until the general condition has received attention.

Treatment of Cases of General Rickets. — The treatment is directed to the improvement and regulation of the nutrition. The medicinal part consists of the administration of phosphorus in some form, usually in that of Thompson's solution,

℞

Phosphori.	gr. i.
Alcohol absol.	m. cccl.
Spr. menth. vir.	m. x.
Glycerine	ad ℥ ii.

Sig.

Three to ten drops. t. i. d.

or else combined with cod-liver oil, and the amount given to a child of two to four years of age is from 1/200 to 1/100 of a grain, and its administration is omitted every two weeks for three or four days, and then resumed in the same dose. In addition to the medicinal treatment, the dietary and hygienic regulation is insisted upon as being of much importance. In some form or other, fat is always prescribed, either as cod-liver oil, cream, or butter. The dietary restrictions consist in the avoidance of all kinds of indigestible and rich food and pastry, of tea and coffee, and the giving in generous quantities of milk, meat juice, eggs, fruit, and, to a moderate extent, carbo-hydrate food. Parents are directed to bathe the children daily, and, except with infants, to rub them down well after the bath. They are also directed to keep them out of doors at all times of the year as much as possible, and are cautioned to carefully regulate their bowels.

These cases are not forced to walk or to sit up, but are kept as much as possible in the recumbent position while the bones are soft. When bone deformity is present, and particularly if in the back, this is insisted upon, and the children are kept on frames, as described under the head of rachitic spine.

SECOND CLASS: RICKETS WITH DEFORMITY OF THE EXTREMITIES.

The majority of these cases are such as come under the head of bow-legs, knock-knees, and flat feet, and their local treatment is considered elsewhere.

THIRD CLASS: RACHITIC SPINES.

A large number of the younger cases of rachitis present the typical deformity of the spine, — the so-called rachitic spine, — which presents a long continuous curve, usually with a point of greater prominence in the lower dorsal region, and most noticeable when the child is sitting with the legs extended, and to a less extent in the standing position. As a rule, the prominence disappears during recumbency, but may not do so entirely. Occasionally, however, a knuckle closely resembling that found in true caries is seen. There are, as a rule, but few symptoms referable to the back, — but manipulation sometimes gives pain, and general tenderness may exist.

Treatment of Rachitic Spine. — As cases presenting this condition are almost invariably found in the earlier stages of rachitis, and therefore show the general symptoms, general treatment directed to the nutrition is always prescribed. In addition to this, the local condition receives attention, and in order to remove all superincumbent weight from the already distorted spine, these children are placed on the rectangular frame, and are kept in this position as long as they present any symptoms which point to the presence of any general rachitic process. The treatment of these cases on the frame does not differ essentially from that in spinal caries, except that absolute fixation is not enforced, but the children are removed to be bathed, and the legs are not bound to the frame. On this they are carried about, and kept out of doors as much as is possible.

ACUTE RACHITIS.

The condition described by some authors as acute rachitis is frequently met with, and occasionally presents pseudo-paralysis. As these cases present no peculiar symptoms or phases of the disease other than those of general rachitis, they need no particular description. The treatment is that of rachitic spine, except that greater attention is paid to the avoidance of all jar and motion.

E. G. B.

BOW-LEGS.

THERE are two broad types of curves recognized, the lateral and the antero-posterior. The former includes all those cases of general lateral curve, whether affecting the tibia alone, or the femur and tibia together, while the latter are seen as sharp and usually angular curves, affecting in the large majority of cases the lower third of the tibia, and occasionally the femur.

The first of these is treated both by apparatus and operation, but the second is seldom, if ever, treated except by operation, and thus must be considered separately under operative treatment.

The Treatment of the Cases of Lateral Bowing.

The treatment of these cases is influenced by the age of the patient, the location and the degree of the curve, and the thoroughness with which the brace treatment can be carried out. It may be divided into: (1) expectant; (2) by apparatus; (3) operative.

The Expectant. — It is well known that there is a strong tendency for this deformity to correct itself during growth, and for this reason a large number of young children are not given local treatment when first seen, unless the degree of the deformity is marked, but attention is given to the diet, hygiene, and general condition of the patient, and the child kept under observation. Tracings and measurements of the curve are taken, and the parents are told to report every two months, or more frequently if treatment of the general condition requires it, and during this time the deformity of the legs is watched, and if it is found to be increasing, or if no improvement is seen, an apparatus is applied.

The tracings are taken in the following manner: The child is made to lie upon a stiff sheet of paper with the feet together, and the outline of the legs and the lower part of the trunk is then drawn. To avoid inaccuracy in the tracing, from the turning of the pencil from the vertical, the following device may be used: To one point of a triangular disc of metal is secured a perpendicular tube, in which is placed a marking pencil, the point of which projects below the surface of the metal. If the disc is held flat on the paper, and the pencil in close apposition to the leg, the tracing thus obtained is an accurate reproduction of the outline of the leg. For further completeness of record, the malleoli are placed in contact, and the distance between the condyles are taken, both with the patient lying and standing. If the level of the condyles shows any departure from the normal, their outline is taken by means of a strip of lead, and recorded. Directions are given for daily massage. The foot and ankle are grasped so that the thumb and forefinger are placed above the malleoli, and pressure is made by the hand rubbing on the outside of the extended leg. By this grasp of the foot no strain comes upon the ligaments of the ankle joint, while that on the knee is not sufficient to cause stretching, but if there is fear of this, counter resistance may be made at this place as well as at the ankle. Children are allowed to move about at will, and if the deformity shows any tendency to increase, apparatus is applied.

Treatment by Apparatus. — This treatment is used in all cases under four or five years, in which thorough attention to the care of the apparatus may be expected from the parents at home, when in such cases the bone is still yielding; but where apparatus treatment will not be faithfully carried out, as occurs frequently in an out-patient clinic, operative treatment is advised.

The usual form of splint consists of a flat strip of steel fastened to the sole of the shoe, and extending upward on the inner side of the leg nearly to the perineum, crossing from this point anteriorly and obliquely across the thigh to a point opposite the trochanter. The apparatus either terminates here by a strap which connects this point to the upper extremity of the inner portion of the upright, or it is joined to a short horizontal band,

APPARATUS FOR BOW-LEGS.

which in case of a double apparatus is fastened to that of the other side, and in cases of single is fastened to a pelvic girdle. There is a joint with a pad at the ankle, and a small plate at the upper end of the inner upright. A broad leather on the outer side of the leg reaching from above the knee to two or three inches below, or to beyond the prominent part of the curve, is then secured to the upright by straps passing behind and in front of the leg, and fastened by buckles on the upright.

The apparatus is applied with the child lying on its back. The lower part is secured by fastening the shoe, and the pelvic band is then strapped to the waist. The broad leather pads are then strapped around the leg, drawing the knee gently against the upright. When the child stands, the tendency of the leg to bow is sufficient to increase the tension to a point that is comfortable, or that the child can endure.

Operative Treatment. — Cases which come to operation are those which are too advanced in age or degree of deformity for the splint or the expectant treatment, or those in which the application of the apparatus cannot receive proper attention. The operation performed consists in breaking the bone and securing the leg in the corrected position. The fracture is made by an osteoclast, except in cases of sharp angular deformity, or of anterior bowing, when an osteotome is often used, or a wedged shape exsection is made. When the osteoclast is used, both tibia and fibula are broken, and if necessary the femur also; but as a rule the fracture of the femur is left to a later period. After the fracture, the legs are placed in a plaster-of-Paris bandage from the foot to the thigh, or if the operation is bi-lateral, or the femur is fractured, the plaster is carried to the waist and a double spica applied. The limb is kept in the cast for from five to six weeks, and if at the end of this time there seems any danger of bowing, splints are fitted and worn for from six months to a year.

E. G. B.

THE cases of knock-knee which come to the Hospital group themselves in a general way into three classes: first, those in which there is but slight deformity, and where the treatment is wholly general; second, those in which apparatus is applied for the direct correction of the deformity; and third, those in which, from the nature or the extent of the deformity, apparatus would not be efficient, and operation is resorted to.

In all cases, whatever is to be the treatment, whether operative or otherwise, special attention is paid to the child's general condition; since almost invariably, coming as the majority of these children do from the foreign poor population, the deformity is associated with more or less nutritive disturbance, either of a simple marasmic nature, or of the true rachitic type. The diet is carefully regulated, and an abundance of fresh air, together with regular bathing, is advised. For medicinal treatment, iron or phosphorus, together with cod-liver oil, is used. If, when the child is first seen, there is besides the knee deformity, evidence of acute rachitis, best shown by epiphyseal pain or tenderness, the constitutional treatment is advised, and at the same time the child is kept in bed, or better, upon a frame, by which it is possible to take the child easily and regularly into the open air. Besides this, if it be deemed necessary, the legs are bandaged to straight splints. This treatment is continued until the evidences of the acute trouble have disappeared, when the apparatus is applied or the operation performed, as may be necessary.

In the first group of cases, those in which apparatus is not necessary, the deformity is often due to flat-foot, or faulty positions in standing as well as to muscular weakness. These cases are much benefited by massage and medical gymnastics, together with salt water bathing. If the arch of the foot has given way this should be supported, and any faulty attitudes rectified.

The cases requiring apparatus — those of the second group, which is by far the largest of the three — are usually seen between the age of one and four years, and the deformity is almost always due, to a greater or less extent, to laxity of the internal lateral ligaments, as well as to some curvature of the bone. The apparatus shown in the figure consists of a steel upright on

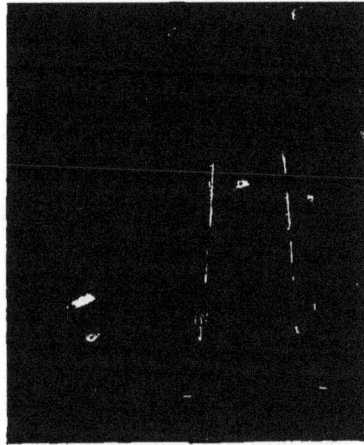

APPARATUS FOR THE CORRECTION OF KNOCK-KNEE.

the outside of the leg, which extends from the top of the great trochanter to the sole of the foot, with a right angle joint at the ankle. This is attached by means of a sole plate to a strong laced shoe. At the upper end a leather strap is attached, which encircles the body, while at the knee, by means of four buckles, two above and two below the joint, is attached the leather pad by which the leg is drawn toward the upright. By regulating these straps the leg can be straightened, little by little, until the deformity has entirely disappeared. There being no joint in the apparatus at the knee, the child of course walks "stiff-legged," and at first is rather awkward, but very soon becomes accustomed to the constrained position.

The apparatus is worn continuously during the day for from eighteen months to two years, being lengthened and readjusted from time to time as the child grows.

When it is evident, either from the extent of the deformity, or because of the undue length of the inner condyle of the femur, that the desired result cannot be brought about by apparatus alone, the child is admitted to the hospital, and the leg straightened by operation. For the day before operation the child is kept in bed with a wet corrosive dressing, 1-3000, applied to the part to be operated upon. The operation is a modification of that known as "Macewen's," or supra-condyloid osteotomy, the difference from the classical operation being that during the division of the bone the leg is abducted and flexed, and rests upon its outer surface. The tubercle of the adductor magnus is carefully made out, and the osteotome is entered one-half inch above this point, being driven directly through the skin without previous incision. After reaching the femur, the osteotome is turned at right angles, and a section made through half the circumference of the bone, care being taken to divide the shell of the bone rather than the cancellated structure. After this partial section the fracture is completed by manual force, the intention being to make an impacted fracture, especially on the inner side. If there is combined with the knock-knee an anterior bowing, or outward twist of the tibia, which is not infrequently the case, a division of the tibia is made at the appropriate point, and this deformity is corrected. Small pads of gauze saturated in sublimate are applied to the wounds, and the leg wrapped in sterile sheet wadding, from the toes to the groin, outside of which a carefully fitted plaster-of-Paris bandage is applied. Both legs are usually operated upon at the same time.

After the operation, as a rule, there is comparatively little bleeding, and not much pain. The temperature may rise on the first or second day to 100° or 101°, but should fall again on the third or fourth day. The children are kept in bed or on a frame for one month; the plasters are then removed, and the use of the legs allowed, so that a week or two later the child should be able to walk. Occasionally it is found that there is marked relaxation of the lateral ligaments, and it is necessary to apply an apparatus temporarily to hold the limbs in their corrected position, the relaxed ligaments eventually becoming strong.

There have been no deaths following this operation, and in only one instance has a serious complication arisen, where an osteomyelitis resulted, which demanded the exposure of the inflamed bone, and the excision of the diseased mass. This resulted in a shortened but useful limb.

In rare cases, where it has been necessary to remove the plaster in the course of the treatment, it has been immediately reapplied. No other method has been used to immobilize the legs, as plaster-of-Paris has been found perfectly satisfactory. As to the age for operation, if the child is below three years, the operation is, as a rule, not performed, although a few cases have been operated upon when but a little over two years old.

In all the cases, operative or non-operative, record

tracings of the legs are taken at the first visit, and then at frequent intervals, so that the condition of the patient can be carefully watched, and any improvement or increase of the deformity can at once be seen.

As to the ultimate results in operative cases, they are good, so far as is known. In 1889, the writer looked up the results of these cases, and was able to get complete records in twenty-eight which had been operated upon for knock-knee or bow-legs, and in all of these, with one exception (a case in which there had been very marked deformity, with some evidences of the rachitic process still being active), the correction of the deformity had been permanent. The cases were all of them seen at least a year and a half after operation.

The result of the experience of the Hospital has been that a few cases of very mild knock-knee can be corrected by attention to the diet, general hygiene, and the correction of the faulty attitude of the patient; that a larger number of cases can be corrected by the application of an outside upright with intelligent supervision of the apparatus by the parents; and that in the cases not amenable to such treatment, operation may be performed with slight danger and great confidence of success.

H. L. B.
J. E. G.

FLAT FOOT.

THE cases of flat foot which are presented for treatment at the Hospital form but a small percentage of the whole number of cases seen, and occur for the most part in young children who are just beginning to walk, or have walked for a short time. The cases of flat foot in the older children do not differ materially from the condition found in young adults, and the treatment given them is practically the same.

The cases which occur in children may be grouped under three heads: (1) cases occurring in flabby and rather poorly developed children, in which there is not only a flattening of the arch, but in which the foot shows more or less tendency to the valgus position, with eversion at the ankle joint. These cases usually show other signs of lack of muscular development and rickets, and require general treatment as well. (2) Cases which occur in children who are strong and otherwise well developed, but who are particularly heavy, and in whom the arch of the foot has never formed or else has broken down, apparently from the disproportion between the weight and the strength. These cases show, as a rule, but little tendency to eversion of the foot, and their general condition does not require special attention. (3) Cases which occur in rachitic children and present, in addition to the general evidences of rickets, the typical flat everted foot.

The imprint of the sole in these cases of young children resembles closely that in the severer cases of flat

foot in adults, in which the whole sole of the foot is in
contact with the ground. This condition is however
rarely associated with prominent symptoms unless accompanied by an eversion of the foot, when easy fatigue
in walking is prominent. The imprint of the foot in these
two conditions is shown in the accompanying cut.

The clinical history of the cases of flat foot occurring
in children shows an absence of prominent symptoms,
and in fact it may be said that symptoms are usually.

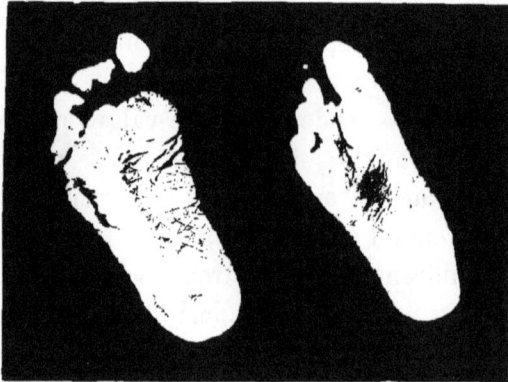

FLAT FOOT.   FLAT FOOT WITH EVERSION.

absent, and the children are brought here on account of
the appearance of the foot, rather than because of symptoms arising from it.

The local treatment is the same for all the types
described. The general treatment need not be considered in this place. The object of the treatment in
all cases is to prevent, as far as possible, eversion of the
foot, to prevent abduction of the foot, and to support the
arch. It is known that the foot in small children may
normally present no arch, and the absence of this is regarded only where eversion or abduction of the foot is
also present.

The methods which are used to correct this deformity are the Thomas shoe, felt sole plates, metal sole plates, and metal plates fitted with uprights. In cases of simple flat foot, felt pads are used and worn on the inside of the shoe. These are made by fitting an insole to the ordinary shoe, and under the arch is placed a pad of saddler's felt, of such shape and size as to give the desired support. In mild cases this is sufficient to support the foot; and as the child becomes stronger and develops, the arch gradually appears, and the foot is soon of normal shape.

In those cases, however, where there is eversion of the foot so that the weight falls to the inside of the normal point, the inside of the shoe is raised after the method of Thomas by a wedge of leather secured to the sole of the boot. As this causes the foot to be held in a position of inversion, the weight falls farther to the outer side and nearer the normal point.

Where more support is needed than is obtained by either the felt plates or by the method of Thomas, a light metal sole plate is fitted and worn in the same manner as in adult cases. The form which is found to give the most satisfaction with young children is a long plate which nearly covers the sole, and is raised on the inside as high as the head of the astragalus.

These are best made over a plaster cast which, in the case of children, is most easily and accurately taken by placing the feet in the plaster, and allowing it to harden with the foot in this position. If this is done while the child is sitting, no weight is borne on the foot. There is, in this way, but little flattening of the arch, and if more pressure than this is desired, the shape of the cast may be changed, and as much pressure as is desired obtained. In young children, this method has been found preferable to that of Whitman, owing to the time required in taking the cast of the whole foot.

In the severer cases there is considerable eversion of the foot as well as a breaking down of the arch, so that the condition is one which may be called splay foot. Obviously something is necessary which will give more support than is obtained by either of the above means, as in these cases it is necessary not only to correct the abduction of the foot and the lowering of the arch, but also the eversion at the ankle joint. Such cases are fitted to the valgus shoe, which is made with a steel or aluminum-bronze sole plate fitted to support, as nearly as possible, the arch of the foot, and which has attached an upright on the outside of the leg. If necessary, the foot may be strapped to the plate, and from the inner edge extends a pad of leather resembling the T bandage, the upper arms of which are made to strap around the foot, and over the upright so that the ankle is held from sagging inward when the weight is placed upon the foot.

If there is contraction of the tendo Achilles, as is frequently found in these cases, an appliance is fitted with a right angle stop joint.

In many cases there will be present a decided eversion of the foot at the ankle joint without an appreciable amount of lowering of the arch. These are frequently treated by an outside upright, furnished with an ankle joint and attached to the sole of the shoe, and to this is added a T strap, as has been described, fastened below to the sole of the shoe in place of the sole plate as in the club-foot shoe, and the upper portion is strapped around the upright, so as to pull the ankle outward.

The third class, or the rachitic cases, present a deformity in which the sole of the foot is nearly flat on the ground, with a decided eversion at the ankle joint and a lowering of the internal malleolus.

These cases present other evidences of rickets and frequently a deformity of the legs, as bow-legs and knock-knees, and require other local treatment as well as general. If other local appliances such as for bow-legs or knock-knees are not required, the foot is supported in the manner already described.

A certain number of cases of flat foot are seen in older children, usually during the period of rapid growth, and in these the subjective symptoms are more or less prominent, and frequently far in excess of the physical signs. There is usually present an eversion at the metatarsal joint, and prominence at the head of the astragalus, but frequently there is little or no lowering of the arch. This condition of the foot is easily shown and recorded by allowing the patient to stand on a piece of smoked paper, which leaves an accurate picture of the impact of the sole. The imprint is then fixed by flooding the smoked surface by any fixative solution or shellac, and serves as a record of future reference. There may be decided complaint of pain on standing or walking, and in many of the cases of long duration, the foot shows a decided stiffness and tenderness, and in some cases a marked degree of flattening of the arch exists.

If the foot shows a painful condition, it is put at rest in the plaster-of-Paris bandage in an over corrected position, and is not used for a period of two or three weeks. If at this time the foot has become flexible and the tenderness and stiffness have disappeared, there is fitted a metal sole plate, and, if necessary, the shoes are raised after

Thomas's method. For a metal plate, the pattern described by Whitman has been found to be useful. A cast of the foot is taken, either by the method advocated by him, or by allowing the patient to stand in plaster, which is then built up along the inner and outer sides of the foot to the required height. The mould obtained from this is cut away beneath the arch to give the desired amount of support, and on this corrected cast the plate is made.

E. G. B.

EMPYEMA.

THE cases of empyema which require operation at the Hospital have usually been under medical treatment. As soon as it becomes clear that there is pus in the pleural cavity, an operation is undertaken for its removal. In those cases where the child is very feeble, a delay of from twenty-four to forty-eight hours is made in order to stimulate the patient. Except for purposes of diagnosis aspiration is not used.

The child is anæsthetized with the greatest care, and, if possible, all strangling is avoided as in several instances irritation from the ether has been sufficient to seriously embarrass respiration, and it has been found necessary to perform the operation under primary anæsthesia; so that in every instance, the surgeon is prepared to open the pleural cavity rapidly should alarming symptoms supervene. The child having been anæsthetized, an incision two inches in length is made in the axillary or post-axillary line, over the seventh or eighth rib and down to the bone, with the strictest antiseptic precautions. The diagnosis is then verified by the introduction of a hollow needle, after which the soft parts, including the periosteum, are carefully separated from the rib for a distance of one inch, and this portion of the bone is excised. During this part of the operation the pleural cavity is not opened, the wound being kept clean. In many instances it has been found safer to avoid the delay and the increased shock, of clearing the rib thoroughly after the primary incision;

17

this is done by rapidly excising the rib with a pair of bone forceps, and opening the pleural cavity at once. This is, of course, followed by a gush of pus. If the patient is strong, the pleural cavity is washed out with sterile water, or a four per cent solution of boracic acid. If shock is especially to be avoided, the bulk of the pus which has accumulated in the pleural cavity is allowed to drain rather than to gush away. Double drainage tubes of large calibre are introduced, secured by safety pins that they may not fall into the pleural cavity, and these are surrounded with a ring of iodoform gauze, and a dressing of sterile absorbent pads is rapidly applied.

In a few instances, Cabot's dressing has been applied. This consists of a piece of mackintosh laid over and overlapping the first layers of the dressing to act as a valve to hasten the expansion of the lung. In the majority of instances however, a roller gauze bandage secures a very large dressing, into which the pus leaks. As soon as the patient recovers from the ether, or the next day, the dressing is changed if it is found saturated with pus or serum, and the cavity is washed out with warm sterile water, and a new dressing is applied. This second dressing is usually kept on for two or three days unless the temperature rises.

In the excision of the rib, there has never been any troublesome bleeding from the intercostal arteries. Pressure alone checks the hemorrhage. A ligature is rarely necessary. It is found that even where the periosteum is not carefully saved, that the bone is reproduced, and at the end of a few years nothing but a scar in the soft parts remains. In some of the cases reported in the accompanying table, a drainage tube was introduced through an intercostal space, but of late the excision of a rib to secure free drainage is done in every case. Occasionally it has been necessary to do a second operation to enlarge the opening into the pleural cavity. In one instance the pressure of the drainage tube on the ribs caused the periosteum to throw out two bridges of new bone, one on either side of the tube, in this manner uniting the sixth and seventh ribs.

The "after treatment" of these cases is of great importance. Whenever the temperature rises, where there is any dyspnoea, or the child loses appetite and strength, the dressings are frequent. The cavity is washed out with warm sterile water, boracic acid, or chlorinated soda wash, one part to sixteen, and the tubes are kept in position as long as there is any discharge, which is usually from four to six weeks. From time to

WELDING OF RIBS FROM PRESENCE OF DRAINAGE TUBE.

time, auscultation and percussion demonstrate the expansion of the lung. At the end of one to three months the case is cured.

Occasionally pus has collected after the too early removal of the tubes. Necrosis of the ribs has never occurred. The children at the end of the third or fourth

TABLE OF CASES OPER–

No.	Date of Operation.	Name.	Age.	Pulmonary Symptoms.	Side.	Operator.	Operation.
1	May 15, 1885.	Daniel Holland.	2 yrs.	For 2 months.	R.	Cabot.	Resection rib under carbolic spray.
2	May 24, 1887.	Mary L. Bremmer.	2½ "	Pneumonia.	L.	Bradford.	Resection 7th rib in post-axillary line. One inch removed.
3	June 8, 1887.	Anna Currie.	4¾ "		L.	"	Incision 7th intercostal space in post-axillary line.
4	Aug. 14, 1887.	George Pfeiffer.	6½ "		R.	Cabot.	Incision 7th intercostal space in post-axillary line.
5	Feb. 1, 1888.	Nellie Slater.	10 "	Pneumonia.	R.	Bradford.	Resection 7th rib in ant.-axillary line.
6	Jan. 26, 1889.	Joseph Sullivan.	5 "	Pleurisy (serous), aspirated three times.	L.	"	Resection 7th rib in post-axillary line.
7	June 10, 1889	Thomas Fitzpatrick.	6 "	For 6 weeks.	L.	"	Resection 7th rib in post-axillary line.
8	July 14, 1889.	Kate Drohan.	7 "	For 1 month.	L.	"	Resection 7th rib in post-axillary line.
9	Nov. 6, 1889.	George E. Bright.	8 "		R.	"	Resection 8th rib in post-axillary line.
10	Jan. 27, 1890.	Joseph Sullivan.	4 "	Pleurisy.	L.	"	Resection 7th rib in mid-axillary line.
11	Jan. 28, 1890.	Samuel Bell.	6 "		R.	Burrell.	Incision 7th intercostal space in post-axillary line.
12	May 5, 1890.	Katie Ward.	8 "	Pneumonia.	L.	Bradford.	Resection 8th rib in post-axillary line.
13	July 1, 1890.	James Fitzpatrick.	2½ "	For 4 weeks.	L.	Lovett.	Resection 8th rib in post-axillary line.
14	Nov. 5, 1890.	Roger C. Elwell.	3½ "	Pneumonia.	L.	Bradford.	Resection 7th rib in line of scapula.
15	Dec. 16, 1890.	Sarah Eldredge.	6 "	For 6 weeks.	L.	"	Resection 7th rib in post-axillary line.
16	Jan. 24, 1891.	May Kennedy.	5 "	Pneumonia.	L.	"	Resection 7th rib between scapula and post-axillary line.
17	Nov. 4, 1891.	Margaret Kelly.	7 "		R.	"	Resection 7th rib in post-axillary line.
18	Nov. 11, 1891.	Helen Leftawitch.	12 "	Pneumonia.	R.	"	Resection 7th rib in post-axillary line.
19	Nov. 16, 1891.	George McGee.	6 "	Pneumonia.	L.	"	Resection 7th rib in post-axillary line.
20	Nov. 14, 1891.	Arthur G. Nelson.	7 "	Pneumonia.		Burrell.	Incision 7th intercostal space in post-axillary line.

ATED ON FOR EMPYEMA.

Dressing.	Tubes Removed.	Duration of Discharge.	Stay at Hospital.	Recurrence.	Ultimate Results.	Remarks.
Irrigation with Hg.Cl₂ 1-15000, 2 drainage tubes, antiseptic dressing.	1st on 2d day. 2d on 8th day.	3–4 mos.	56 days.	Yes.	Well.	Recurred 11 months after closing of sinus; dis. 6 weeks, leaving no abnormal signs ; not reported since.
Drainage tube, antiseptic dressing.	1 month.	1 mo.	32 days.	No.	Well.	Reported Aug. 23, 1893 ; perfectly well.
Irrigation with 1-20 chlor. soda, Cabot dressing, later with Hg.Cl₂ 1-15000.	10 days.	21 days.	23 days.		Well.	Not reported since.
Irrigation with 1-20 chlor. soda, Cabot dressing, later Hg.Cl₂ 1-15000.	23 days.	38 days.	40 days.		Well.	Not reported since.
1 drainage tube, irrigation with Hg.Cl₂ 1-15000, antiseptic dressing.	20 days.	3 mos.	2 mos.	No.	Well.	Aug. 23, 1893 ; perfectly well. Sl. dimin. resp. over lower r. back ; lat. curvature giving marked deformity ; heart nor. situated ; lungs expanded.
Cabot dressing, irrigation with Hg.Cl₂ 1-10000.	1st on 8th day, 3d on 11th day.	25 days.	32 days.	Yes.	Well.	Recurred 5 months after closing ; 2d. oper. ; incision over sinus tubes ; dis. stopped one month ; not since reported.
Irrigation with sterilized water, Cabot dressing.	11 days.		26 days.		Unknown.	Sl. dis. on leaving ; not since reported ; child in good condition.
Irrigation with 1-20 chlor. soda, 1 drainage tube.	8 days.		24 days.		Unknown.	Sl. dis. on leaving ; not since reported ; child in good condition.
Irrigation with Hg.Cl₂ 1-10000, chlor. soda, Cabot dressing.	2 days.	32 days.	65 days.	No.	Well.	Aug. 22, '93, w. developed ; constitutional symptoms and pulmonary signs of lung tuberculosis ?
Irrigation with sterilized water, 2 drainage tubes, baked dressing.	7 days.	20 days.	23 days.		Well.	Not since reported.
Irrigation with Hg.Cl₂ 1-10000, hydro - napthol, Cabot dressing.	18 days.	28 days.	34 days.		Well.	Lung in good condition ; not reported since.
Irrigation, Cabot dressing.	5 days.	68 days.	14 days.	No.	Well.	Dec. 9, 1890, perfectly well ; sl. rel. dullness r. lower back ; weak resp. murmur.
Irrigation with Hg.Cl₂ 1-10000, Cabot dressing.	10 days.	35 days.	38 days.	No.	Well.	Aug. 22, 1893, l. chest perfectly normal ; rugged child.
Irrigation with Hg.Cl₂ 1-10000, Cabot dressing.	10 days.	16-17 mos.	86 days.		Dead.	Complic. lobar pneumonia l. side ; recur. 13 mos. after 1st op. ; retrac. 6, 7, and 8th ribs post-ax. ; died exhaustion.
Irrigation with sterilized water, Cabot dressing.	24 days.	3 years.	1 mo.	Yes.	Well.	Aug. 23, 1893, good gen. con. ; l. chest retracted ; heart displaced, lat. curv. ; l. lung, fair resp. over it ; no rales.
Irrigation with Hg.Cl₂ 1-10000, Cabot dressing.	8 days.	67 days.	29 days.		Well.	Not reported since ; discharge stopped.
Irrigation with Hg.Cl₂ 1-10000, Cabot dressing.	2 days.	11 mos.	28 days.	Yes.	Well.	2d op. 3 mos. after 1st ; reseo. 7th rib ; Cabot dressing ; tube removed 11th day ; sinus closed in 11 mos. from 1st operation ; not reported since.
Irrigation with Hg.Cl₂ 1-10000, Cabot dressing.	14 days.	23 days.	24 days.		Well.	Not reported since.
Irrigation with Hg.Cl₂ 1-10000, Cabot dressing.	15 days.	57 days.	50 days.	Yes.	Unknown.	2 weeks after closing, reopened ; not reported since.
Irrigation with sterilized water, Cabot dressing, later with chlor. soda 1-16.	22 days.			Yes.	Well.	Second operation 5 mos. after 1st ; sinus closed and reopened ; 6th and 7th ribs resected in post-axillary line ; tube removed 25th day ; discharged 45 days, reopened again 3 weeks ago, very sl. discharge ; Aug. 23, 1893, sound wound. sinus closed, apparently no fluid in chest, considerable lat. curvature r't. side, lung fairly expanded.

TABLE OF CASES OPER-

No.	Date of Operation.	Name.	Age.	Pulmonary Symptoms.	Side.	Operator.	Operation.
21	May 5, 1891.	Fred Rugler.	3½ yrs.		L.	Bradford.	Incision in old sinus, one inch inside nipple line.
22	May 19, 1891.	Adeline V. Willis.	4 "	Pleurisy.	L.	"	Resection 8th rib in post-axillary line.
23	July 1, 1891.	James Sweeney.	6½ "	Pulmonary symptoms, 5 weeks.	L.	Cushing.	Resection 8th rib in post-axillary line.
24	Feb. 9, 1892.	Edward Drool.	3 "	Pneumonia.	L.	Bradford.	Resection 5th rib in ant.-axillary line ; 7th rib in post-axillary line.
25	Apr. 27, 1892	Elsie Rumstner.	7 "	Pneumonia and pleurisy, for 7 weeks.	L.	"	Resection 8th rib in scapula line.
26	Apr. 27, 1892.	Peter Kittredge.	3 "	For 7 weeks.	L.	"	Resection 5th rib at inf. angle scapula.
27	Jan. 27, 1892.	Willie Quinn.	6 "	Unknown. Entered with spontaneous opening.	L.	Cushing.	Resection 7th rib over sinus at angle of scapula.
28	Nov. 27, 1892.	Stella Kimball.	5 "	For 5 weeks.	L.	Bradford.	Resection 7th rib in axillary line.
29	May 10, 1892.	John Muldoon.	10 "	Pneumonia.	R.	"	Resection 7th rib in axillary line.
30	May 14, 1892.	Annie Cohen.	4 "	Pneumonia.	L.	Burrell.	Resection 7th and 8th ribs in post-axillary line.
31	June 27, 1892.	Elizabeth Libbey.	4 "		R.	Bradford.	Resection 8th rib in post-axillary line.
32	July 21, 1892.	Alice Carroll.	3 "	Pneumonia complicated with pericarditis	L.	Brackett.	Resection 6th or 7th rib in post-axillary line.
33	July 21, 1892.	Helen Torrey.	7 mos.		L.	Cushing.	Resection 5th rib between post-axillary and scapula line.
34	Aug. 27, 1892.	Carl Zepliffe.	2½ yrs.		R.	"	Resection 8th rib in post-axillary line.
35	Aug. 23, 1892.	Helen Devine.	2 "		R.	Burrell.	Resection 8th rib in scapula line.
36	Aug. 25, 1892.	Bessie Dodge.	10 "	Pneumonia.	L.	Cushing.	Resection 8th rib in scapula line.
37	May 11, 1893.	Willie Barrett.	11 "	For 1 month.	L.	Lovett.	Resection 6th rib in post-axillary line.
38	May 28, 1893.	John Donahue.	1¾ "	Pneumonia.	R.	Bradford.	Resection 7th rib in post-axillary line.
39	May 8, 1893.	Alice Kennedy.	6⅔ "	Pneumonia.	L.	"	Resection 7th rib in post-axillary line.
40	June 18, 1893.	Charles Reynolds.	10 "	Pneumonia.	L.	Burrell.	Resection 8th rib in mid-axillary line.
41	June 22, 1893.	Philomene Curnhan.	3 "	Pneumonia.	R.	"	Resection 8th rib in mid-axillary line.
42	July 12, 1893.	George Ludgy.	2½ "	For 3 weeks.	R.	Bradford.	Resection 7th rib in post-axillary line.

ATED ON FOR EMPYEMA.

Dressing.	Tubes Removed.	Duration of Discharge.	Stay at Hospital.	Recurrence.	Ultimate Results.	Remarks.
Irrigation, Cabot dressing.			2 days.		Dead.	Child in wretched condition on entrance; died 24 hours after operation; shock.
Irrigation with Hg.Cl₂ 1-10000, antiseptic dressing.	2 days.	162 days.	72 days.	No.	Well.	Aug. 22, 1893, general condition excellent: examination, heart normal position, good resp., murmur over lung.
Irrigation Hg.Cl₂1-10000, Cabot dressing.	21 days.	32 days.	32 days.		Well.	At discharge, lung expanding well, good respiration; not reported since
Cabot dressing, irrigation with Hg.Cl₂ 1-20000.	1st on 3d day, 2d on 31st day.	9-10 mos.	45 days.	No.	Well.	Aug. 20, 1893, no recurrence; perfectly well; sinus had opened spontaneously 8 months before; sl. lateral curvature.
Irrigated with sterilized water, Cabot dressing, later with Hg.Cl₂ 1-10000.		37 days.	16 days.	No.	Well.	Aug. 19, 1893, sl. rel. dullness lower rt. back, no lateral curvature.
Irrigated with sterilized water, Cabot dressing, Hg.Cl₂ 1-20000.	1 month.	10 weeks.	41 days.	No.	Well.	Aug. 19, 1893, perfectly well.
Irrigation Hg.Cl₂ 1-20000, Cabot dressing, tube and iodoform gauze.	Last one 10 mos.	9-10 mos.	5 mo.	No.	Well.	Aug. 24, 1893, boy rugged, perfectly well; fair respiration over lung, sl. retraction.
Irrigation with Hg.Cl₂ 1-20000, Cabot dressing.	20 days.	34 days.	34 days.	No.	Well.	Aug. 20, 1893, child perfectly well.
Irrigation with Hg.Cl₂ 1-15000, Cabot dressing.	36 days.	67 days.	39 days.	No	Well.	Aug. 28, 1893, child perfectly well.
2 drainage tubes, irrigation with Hg.Cl₂ 1-10000.	59 days.		60 days.			Sl. discharge on leaving hospital; no later report.
Irrigated with Hg.Cl₂ 1-20000, Cabot dressing.	14 days, later iodof. wick.	33 days.	19 days.	No.	Well.	Aug. 22, 1893, heart normal position; no lat. curvature; rel. dimin. resp. and dullness of lower back.
Irrigated with Hg.Cl₂ 1-20000, drainage tubes.			11 days.			Removed against advice, in fair condition; no report since.
Irrigated with Hg.Cl₂ 1-10000, Cabot dressing.	25 days.	36 days.	46 days.	Yes.	Well.	Returned 3-4 mos. after 1st operation for carious rib at point of resection, curetted, sinus healed completely in 15 days; no report since.
Irrigation, aseptic dressing.	30 days.	41 days.	42 days.	No.	Well.	Aug 27, 1893, child reported to be in perfect health.
Irrigated with Hg.Cl₂ 1-10000, 2 drainage tubes.			25 days.		Dead.	Died of tubercular meningitis? convulsions, vomiting, etc.
Irrigated with Hg.Cl₂, 2 drainage tubes.	25 days.	2-3 mos.	30 days.	Yes.		Opened spontaneously 5 mos. after closing; drainage and tube after dilation of sinus; July 1. 1893, feverish, weak, considerable discharge.
Irrigated with Hg.Cl₂ 1-10000, drainage tube, Cabot dressing, later peroxide hydrogen.	9 days.	26 days.	43 days.	No.	Well.	Aug. 1893, well, lung normal.
Irrigated with Hg.Cl₂ 1-10000, 2 drainage tubes.	22 days.		38 days.			Slight discharge on leaving; no report later.
Irrigated with 4 per cent solution boracic acid, 2 drainage tubes, antiseptic dressing.		45 days.	55 days.		Well.	Gen. condition excellent, lung nearly normal, slight lateral curvature.
Irrigated with Hg. Cl₂ 1-10000, 2 tubes, later with 4 per cent solution boracic acid, later wick of iodoform gauze.	7th day.	57 days.	60 days.	No.		Lung fully expanded, heart normally situated, no lateral curvature, general condition excellent.
Irrigated with Hg.Cl₂ 1-10000, 2 drainage tubes.	1st on 2d day, 2d on 4th day.	10 weeks.	71 days.		Well.	Lung in excellent condition, general condition good.
Irrigated with 4 per cent boracic acid, 2 tubes.			17 days.			Sent to convalescent home in poor condition; Aug. 30, 1893. general condition much improved, slight discharge.

day begin to take solid food, and often at the end of a week eat voraciously. They rapidly increase in weight and strength, and frequently in six weeks gain ten or fifteen pounds.

The above table of forty-two cases is interesting and instructive, but it cannot be used for statistical purposes, as cases of empyema vary so greatly in their course after operation, that each case is a study in itself.

H. L. B.

SPASTIC PARALYSIS.

FOR purposes of treatment this affection may be divided
into three grades:

I. Light Cases; II. Medium Cases; III. Severe
Cases.

I. LIGHT CASES. — Those in which the rigidity of the
muscles has not reached such a degree as to cause
marked contraction, and can readily be overcome by
the use of slight force. The power of voluntary movement is retained to a considerable extent, and all
motions can be performed, although some may be very
weak.

For these the treatment employed is directed to the
strengthening of the muscles, and by strengthening
those most paralyzed to enable them to overcome not
only their own weakness but also the contraction or
resistance of their opponents.

Three forms of treatment are used for this purpose:

1st, and most important is that by *electricity*. The
Faradic current is applied to the affected muscles, one
pole being placed directly upon the muscles to be treated,
the other, or indifferent pole being held on the small of
the back if the child is too young to hold it in its hand.
The strength of the current must be graduated according
to the sensation of the patient, so far as it can be ascertained, and, if not painful, a current strong enough to
cause contraction in the muscles should be used.

2d. *Massage.* — Something can be accomplished in the way of strengthening the muscles by telling the mother or relatives of the patients to rub the paralyzed limbs, and by showing them once or twice how this should be done. But this does not in our opinion compare in efficiency with real massage as performed by a skilled masseuse. On this account we have lately employed a trained masseuse at the hospital with excellent results.

3d. *Gymnastics.* — This treatment is carried out sometimes in the hospital, and sometimes in gymnasiums to which the patients are sent with directions that they should exercise in special ways, or again they may be instructed in exercises which they are required to practise at home. Forced movements may sometimes be used either in connection with or separately from the massage.

II. MEDIUM CASES. — Those in which the contractions are so marked as to form an essential portion of the hindrance to movement, and yet such as may with time and patience be gradually overcome. Speaking broadly, this applies to all contractions not sufficiently severe to require surgical treatment to lengthen the muscle.

1st. In the milder cases of this group the treatment would be the same as in Group I.; but there are many in which the treatment of the paralysis alone is not sufficient, and something must be done directly to relieve the contraction. This may be in part accomplished by passive motion and massage, especially the latter, which must then be systematically applied by the physician himself, or under his supervision.

2d. *Electricity.* — In cases of such severity that Faradism alone seems of little benefit, we use as a rule a combination of the Galvanic and Faradic currents. We apply the Galvanic current to the contracted muscles, and the Faradic to their opponents, the weaker muscles, using both currents at the same time. This has proved of much service in certain cases, especially in spastic paraplegia.

3d. Apparatus is often advisable in this class of cases, both to correct deformity and to aid the paralyzed muscles.

III. SEVERE CASES. — Those in which the contractions are so firm that they cannot readily be overcome by the methods already described. In such cases deformities caused mechanically by the muscular contractions are the rule. These contractions must be overcome before any other treatment will be efficient, and for this purpose we have recourse to surgery. I have in another part of this report [1] entered more in detail into the methods employed in operating on these cases, and have described at length the results attained. We find that these operations are, if properly conducted, almost invariably successful, — (a) as regards the operation itself, (b) as regards relief of the contraction. On the other hand, operation can only relieve the contractions and cure the deformities caused by them, while it has little or no influence on the causative paralysis. If anything, the weakness of the limbs is rather increased than diminished when the firm support afforded by the contracted muscles is removed. Operative treatment is therefore only the first step in the cure of the more severe cases, but an inevitable first step if the treatment is to be efficient. After operation these cases are brought as regards treatment into Group I. The spasms or contractions being relieved, the paralysis alone remains to be dealt with. They are then treated

[1] See page 327. Tenotomy in Spastic Paralysis, by William N. Bullard.

with Faradism, massage, and, as they become better, by exercises properly chosen and applied.

COMPLICATIONS. — Besides the deformities which are caused mechanically by the contraction of the muscles, and which are usually relieved by overcoming the contraction, there are two groups of complications of great importance which occasionally occur.

The first is feeble-mindedness or idiocy. Some years ago it was considered inadvisable to operate in cases of idiocy, on the ground that even were the operation successful, the patient would not be able to learn how to use his limbs to any extent, and that hence the results were unlikely to be satisfactory. However, this view has of late been somewhat modified. In the first place it is difficult always to determine the exact degree of feeble-mindedness existing in an infant or young child, and as weakminded children increase in age, the mental development may also increase, at least sufficiently to enable the child to use its limbs. Secondly, the operation itself seems in many cases to have a beneficial effect upon the mental development. The change of surroundings, the new impressions necessarily forced upon the child, often seem to act as a mental stimulus, and the possibility of new movements of the limbs, and the pleasure thus obtain d, seem to cause a partial corresponding mental development. There are few or no cases of spastic paralysis presented at the hospital in which I should consider feeble-mindedness a bar to operation, although it might naturally influence our prognosis as to the amount of use of the limbs ultimately to be obtained.

The second complication with which we have to deal is involuntary movements. The practice has hitherto been, when movements of this class existed, to consider them as contra-indications to operation. The value of operation in restraining these movements has not yet been shown. The removal of deformities and contractions can less readily be accomplished in the constant presence of these movements, and even when the contractions have been cured the movements still impede the voluntary actions of the limb and impair its usefulness to a marked degree. In most cases where congenital choreic or athetotic movements exist a marked ataxia also prevails, and this must of itself always be considered as unfavorable to the final result. No form of operation has yet been found which has proved of much avail in congenitally ataxic limbs.

The operative measures used in spastic paralysis are either tenotomy or

SPASTIC PARALYSIS.

myotomy. No attempt is made here to explain the rationale of the operation, but there is no doubt as to its efficacy in certain cases.

The operations which have been found useful have been simple tenotomy of the tendo Achilles, with or without open division of the hamstring muscles. Originally, tenotomy of these muscles was done; but it was

found that open incision was more thorough. The
patient is placed prone, and an oblique incision is made
across the popliteal space. As the contracted tendons
come into view, they are divided thoroughly; the skin
is then sewn, and the limb put up in a corrected position and fixed by a plaster-of-Paris bandage. Both
the tendo Achilles and the tendons of the hamstring
muscles are divided when both are contracted. In a
few instances, myotomy through an open incision over
the adductor muscles has been done. Myotomy in
spastic paralysis of the hand was done in two cases, —
with some benefit, but with much less than in contraction of the lower extremities.

For ten days after the operation, the patient's limb is
fixed by means of a stiff bandage. After this, an apparatus is applied and worn for a few months.

The apparatus for spastic equinus is similar to that in
use for ordinary equinus.

The patient is allowed to walk, ordinarily, in about two
weeks. Where the hamstring muscles are also divided,
in addition to the apparatus for equinus, steel rods connected by a steel band at the top are fitted, which extend
the whole length of the leg and thigh. After the operation has been entirely recovered from, massage and electricity are resumed.

E. H. B.
W. N. B.

TORTICOLLIS.

THERE are three varieties of this affection which present themselves for treatment. First, that dependent upon caries of the spine; second, that due to congenital muscular spasm; and third, the so-called posterior torticollis form, due to some nerve lesion. The acute form rarely comes to a hospital for treatment.

Wry neck, attendant on disease of the cervical spine, will not be considered here, as its treatment is that of caries of the spine.

The cases of congenital wry neck have always been treated by operation. The division of the sterno-cleido-mastoid, and correction of the deformity, are followed by a proper appliance for holding the head in its corrected position. The treatment divides itself, therefore, into three steps: first, operative; second, corrective; and third, mechanical. Both the sub-cutaneous division and the open incision of the attachment of the sterno-cleido-mastoid have been employed. The open incision presents, however, many advantages, being much safer and more thorough. It has for these reasons generally been adopted.

With sub-cutaneous division of the attachment of the sterno-cleido-mastoid muscle, it is comparatively easy to divide the firm bands of fascia which are superficial; but the deeper fibres are often undivided.

An open incision heals readily if the wound is aseptic, and leaves but a slight scar. In the cases which have been operated on, a transverse incision has been made. The skin is pulled down, and a cut made directly over the clavicle. After the tension on the skin is removed, it retracts, and the opening will be found to lie slightly above the clavicle. The attachments of the muscles are thoroughly divided, and any opposing bands of fasciae are cut.

A vertical incision, parallel to the fibres of the sterno-cleido-mastoid, gives a less noticeable scar than a transverse incision in the skin; but it has the disadvantage of not opening the operative field so satisfactorily unless a long incision is made. In one instance the internal jugular vein was wounded in an attempt to divide contracted fasciae beneath the sterno-cleido-mastoid. It became necessary to tie the vein; which was successfully done. In this case the accident occurred in an attempt at unnecessary thoroughness in dividing some portion of the deep cervical fascia. The deep fascia was raised between two forceps and a contracted band was divided carefully. It was subsequently found, that the internal jugular vein lay directly under the deep fascia, and had therefore been cut. This gave rise to no trouble in healing, and the result was satisfactory; but the complication of wounding the vein is one which should be guarded against, and which can be readily avoided if care is exercised. The wound after incision is sewn up, and carefully dressed; it ordinarily heals in a week. Where sub-cutaneous division is employed, and some of the contracted fibres are overlooked, mechanical correction becomes at times extremely difficult.

Other fibres than the sterno-cleido-mastoid are occasionally divided. In one case the scalenus anticus was in part divided.

After the division of the contracted muscular fibres, the head should be forcibly straightened by the hands of the operator. The corrected position can be maintained by the application of a plaster-of-Paris bandage over the trunk, neck, and head. In two instances this was done, but although it enables the patient to go about on the day following the operation, it is cumbersome and does not permit any further stretching of the muscles. For

BED TREATMENT AFTER OPERATION.

this reason in the majority of instances the patient has been placed, immediately after the operation, in a recumbent position upon a frame, and correction has been secured by the use of adhesive plaster strips applied to the face and attached by straps to weights. The accompanying illustration indicates in what way these are applied. It will be found that this occasions but little discomfort, and if worn for a week or ten days will complete correction, and will overcorrect. At the end of a week or ten days, a retention appliance should be furnished the patient, who is allowed to go about freely, using in some instances of obstinate torticollis, the strap at night. The appliance is shown in the accompanying illustration, and is lighter and more easily managed than those usually recommended in books on orthopedic surgery. Ordinarily, it is necessary that this appliance

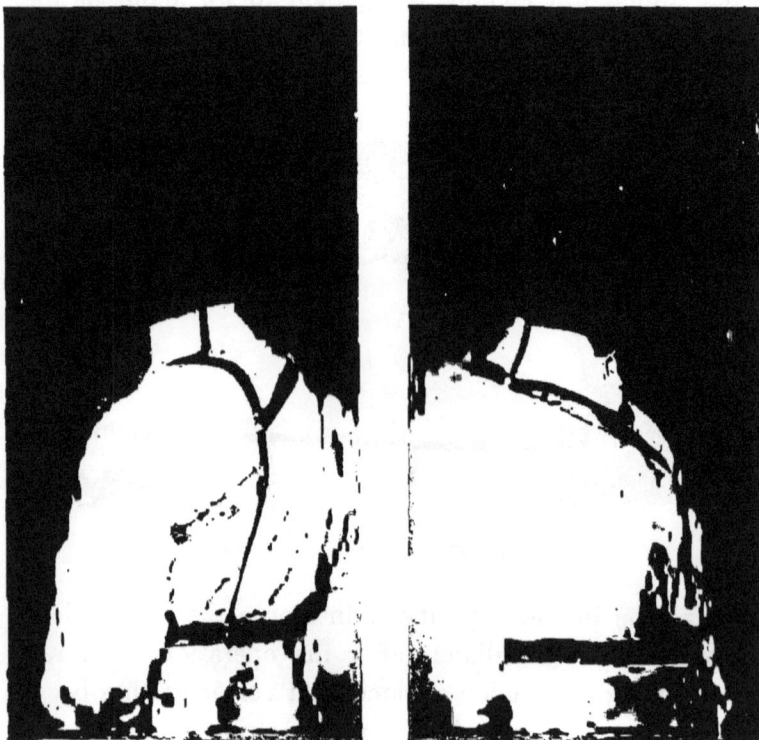

RETENTION APPLIANCE FOR TORTICOLLIS.

be worn for five or six months, after which the cure is complete. In none of the cases have any relapses taken place; and if the division of the contracted tissues is thorough, and the retention appliance properly adjusted, and worn a sufficiently long time, a permanent cure can be expected.

In two patients correction, without division of the sterno-cleido-mastoid or other muscles, was attempted by the use of correcting straps and frame. This method, however, cannot be warranted to yield satisfactory results, although it is probable that if carried out sufficiently long, correction would be possible in young children.

Three cases of torticollis were children in one family. No cause could be found for this, and no hereditary predisposition existed, and the deformity was not known to have occurred in any of the earlier generations. In one instance the deformity appeared to be due to a rupture of the sterno-cleido-mastoid at birth. The patient was an infant a week old, and was brought to the hospital with well marked torticollis; on examination, the sterno-cleido-mastoid was found to be tense and contracted, except in the anterior portion, where a well marked rupture could be felt; at this point there was swelling and ecchymosis. The attendant stated that the birth had been a difficult one. The subsequent history of the case is not known.

Three cases of what has been termed posterior torticollis have been seen at the hospital, — i. e. deformity with contraction not of the anterior muscles of the neck, but of the posterior group. Here tenotomy is not indicated, but where the deformity is liable to be permanent, and is not due to caries of the spine, correction by manual force under an anaesthetic is indicated. The trunk is steadied by an assistant, and the head twisted with care but with considerable force. Adhesions will be felt to give way. The head is then fastened in an over-corrected position, by means of a plaster-of-Paris bandage enclosing the head and thorax; this is worn for several weeks, and is followed by a retention appliance which is worn for six months.   In one instance of this sort, a distortion of three years standing was permanently corrected.

No case of intermittent torticollis is recorded at the hospital.

E. H. B.

HYDROCELE.

HYDROCELE has been treated exclusively in the out-department of the hospital, no case, so far as the writer knows, having been admitted to the wards for operation. No tabulation of the cases as taken from the hospital reports would be accurate, because, where hydrocele and hernia exist together, the cases are generally classified as the latter. It is interesting to see, however, how few hydroceles appear when one contrasts them with cases of hernia. For instance, in the last four years, this table shows the proportion.

	Inguinal hernia. (Males.)	Hydrocele.
1888	36	11
1889	43	10
1890	45	3
1891	50	11
1892	68	13
	242	48

Cases of hydrocele where hernia co-exists are generally slight, and have needed no especial measures beyond the treatment of the hernia. The causes of the cases of simple hydrocele have been obscure in general, although a small proportion have been clearly traumatic. There have been two or three cases of haematoma of the scrotum, which have subsided without active treatment.

Thirty cases of hydrocele, occurring previous to October, 1889, were analyzed in 1890,[1] and they serve

[1] Boston Medical and Surgical Journal, Dec. 18, 1890.

to show the character of the cases applying. Of these thirty cases, nine were double, five proved to be encysted hydrocele of the cord, while the other cases were examples of simple scrotal hydrocele. Cases of slight hydrocele accompanying hernia were not included in this report.

The treatment of hydrocele has been by simple tapping. In recent cases where a traumatic origin was likely, a purely expectant treatment has been followed, and in the thirty cases analyzed, seven were treated in this way, and four recovered.

Of ten cases from the same group which were tapped, nine were permanently cured, the unsuccessful cases in both groups being those where there was communication with the abdominal cavity, which could be demonstrated by the reduction of the hydrocele. These analyzed cases have merely confirmed the general experience, which is that simple tapping (perhaps once or twice repeated) will cure nearly all cases of hydrocele in children where the fluid is not reducible into the abdomen, and that cases of traumatic hydrocele will, if recent, be absorbed spontaneously.

Encysted hydrocele of the cord has, as a rule, been cured by one tapping.

The scrotum is washed with corrosive sublimate solution before tapping. The tapping is done by an ordinary hypodermic needle, attached to a "dental syringe," which is much larger than an ordinary hypodermic syringe. There has been no accident and no infection in any of the cases treated.

R. W. L.

THERE have been twelve cases of spina bifida treated at the hospital. Spontaneous cures have occurred in two cases after the rupture of the sac and shrinking down of the tumor by cicatrization. No success can be reported from excision of the sac. In some instances death has occurred from the shock of the operation, or from the gradual leaking away on the third or fourth day of the cerebro-spinal fluid. The injection of Morton's fluid has been equally unsuccessful; but no deaths are to be reported from its use.

In one case an unsuccessful attempt was made to cure a case by the following method, and as it is applicable to small meningoceles, it is worthy of record: An elliptical incision was made at the base of a large spina bifida in the lower dorsal region. The incision was carried through the skin, and flaps were formed from the loins of the child. The fluid was then drawn from the cyst by an aspirating needle, and the flaps of the skin of the loin were sewed together over the collapsed cyst. There was very little shock from the operation. The cyst, however, gradually refilled with fluid, and finally separated the flaps, and the tumor reappeared in its original size.

The treatment of this disease is at present unsatisfactory. A shield is constructed to relieve pressure and prevent chafing, unless the meningocele will inevitably rupture. In the few cases where the cyst is small, the walls thick, and the opening into the spinal canal not large, Morton's fluid is used. When, however, the sac threatens to burst, and sufficient sound tissue can be

SPONTANEOUS CURE OF SPINA BIFIDA.

found to form flaps, an excision of the sac and isolation of the nerve fibres which are implanted on its inner surface is undertaken. The accompanying cut shows the spontaneous cure of a spina bifida.

H. L. B.

DISEASE OF THE PELVIS.

PRIMARY disease of the pelvis is rare among children, and is usually sacro-iliac disease. Caries of the pelvis, which is secondary to acetabular hip disease, or to low lumbar caries, will not be considered here. There has been one case of apparent primary disease of the inner surface of the pelvis, which resulted in abscess formation pointing in the iliac fossa. There were present no reflex joint symptoms, and relief was given by the evacuation of the pus, by the same operation as for psoas abscess. The few cases of sacro-iliac disease which occur in children present symptoms closely resembling those of low lumbar caries, but have so far all shown symptoms which point to irritation of one or the other hip, with pain extending down the leg, and a diagnosis has been aided by the local swelling over the sacro-iliac synchondrosis. The treatment of these cases has been rest in bed on the rectangular frame, and, later, the application of either permanent or movable plaster-of-Paris jackets. The time of treatment does not vary essentially from that of lumbar caries, and judging by the cases which have so far been treated, the prognosis is about the same.

E. G. B.

TUBERCULAR PERITONITIS.

SEVERAL cases at the hospital have undergone laparotomy for the treatment of tubercular peritonitis.
In all of these, tubercular nodules were found in the peri-
toneum. In one, a culture was made to verify the diagnosis. In three cases, there was suppuration. In one,
there was but little fluid, but the large glands of the
mesentery formed a mass in the abdominal cavity. The
three suppurative cases made an excellent recovery, but
one died a year later with symptoms of tubercular meningitis, the abdominal enlargement having entirely disappeared, and the wound from the laparotomy having
healed. In the three suppurative cases, a drainage
tube was left in for several weeks, and the cavity
was repeatedly washed out. In one case iodoform gauze
was placed in the wound; but as imperfect drainage
took place, an elevation of temperature occurred for
two or three days; a rubber tube surrounded by iodoform gauze was substituted, and retained for a week.
In the cases with serous fluid, the wound was sewed
up after operation, and healed by first intention. In
one case, after the abdominal wound had healed, considerable fluid remained for some time in the abdominal cavity, as ascertained by palpation; but it was
absorbed subsequently, and the patient made a complete recovery, reporting a year afterwards in perfect
health. In the cases of abdominal enlargement with
tubercular glands in the mesentery, and but little fluid

in the abdominal cavity, the operation did not benefit
the patient. In all the other cases improvement followed, and in the serous cases, recovery. This was
also true of one of the suppurative cases, and in the
second, at present under treatment, recovery seems
probable.

E. H. B.

PROLAPSE OF THE RECTUM.

A FEW cases have been admitted to the hospital to be operated upon for prolapse of the rectum. The treatment has been confinement to bed, cold water enemata, and the wearing of a pad kept in place by a T bandage. Astringent suppositories have also been used, and astringent enemata. The rectum in many cases has been found congested and covered with granulations. In several instances, however, no benefit has followed the temporizing treatment, and, under these circumstances, the practice has been to anaesthetize the patient, stretch the sphincter, and cauterize with a Paquelin cautery, the walls of the rectum for a distance of an inch above the opening. The cauterization is linear, the point being applied around the whole circumference of the stretched rectum, in lines at distances of a quarter of an inch apart, burning through the mucus and touching the submucus tissue. Benefit always follows the operation, and a certain number of cases have been apparently cured.

E. H. B.

CONGENITAL DISPLACEMENT OF
THE HIP.

THE mechanical treatment of congenital dislocation has been thoroughly tried in one patient at the hospital, according to the method recommended by Dr. Buckminster Brown. The patient was a girl of three years, with double congenital dislocation. Traction was applied with the limbs straight, and the patient kept recumbent. After two months it was found that the head of the trochanter, which had been an inch above its normal position, could be brought down to its normal place by moderate traction; but it did not remain when traction was relaxed. Daily motion was allowed, with the limbs retained in their proper position by straps, and, at the end of a year, the head of the bone appeared to be retained in place, or nearly so. A year later an ischiatic crutch was applied to each limb so arranged that the whole weight fell upon the crutch and not upon the patient's legs; this was jointed at the knee, and the patient was allowed to go about with this after the third year. The child wore the apparatus for four years, and continues to at present, although it is allowed to walk a part of the time without any apparatus. At the present time the top of the trochanter is half an inch lower than at the beginning of treatment, but still half an inch too high. There is a noticeable improvement in gait and in lordosis; but it is too early to report the ultimate result of the case.

The writers have seen a case which five years before had been thoroughly cured to all appearances, and yet which relapsed after puberty into a condition nearly as bad as at first. The ultimate result in these cases would leave doubt whether this treatment can be expected to be satisfactory.

The operative treatment of congenital dislocation has been tried in five cases, — in one unsuccessfully, in four successfully, although one of these has since died of an acute infectious disease. The method used has been a modification of Hoffa's operation for the reduction of congenital displacement.

The first attempt which was made, and which was unsuccessful, was upon a girl fifteen years of age. Although the head of the bone was replaced in an acetabulum which was chiselled out, it eventually slipped out. The resistance of the soft parts, and the difficulties of this operation, led the surgeons to select younger patients, and now the operation is advised in selected cases on children between two and six years of age. The operation is, of course, possible in older children, but is more difficult.

An incision is made which lays open the posterior capsule of the joint, this is opened and the head of the bone exposed. The soft parts are raised from the great trochanter, and the head of the bone is displaced outwards, a search is then made with the forefinger for the acetabulum. This is deepened by a curette or burr run by a surgical engine, until it is of sufficient size to receive the head. In two instances, it was found that the resistance to the reduction of the head of the femur was due to the outer fibres of the Y ligament as they pass from the anterior inferior spinous process of the ilium to the neck of the bone, and a tenotomy of this resisting band allowed the head to be manipulated into position. If necessary, the abductors of the thigh, and such other resisting structures as interfere with the replacement of the head of the bone in the re-formed acetabulum, are cut. The wound is then sewed up, and the child is placed in a plaster-of-Paris spica bandage, and, unless there is some severe pain or marked elevation of temperature, the wound is not disturbed for a fortnight, when a window is cut in the plaster of Paris and the wound is dressed. Combined with the plaster-of-Paris bandage, adhesive plaster is used to gain extension on the limb. At the end of six or eight weeks, the plasters are removed, and, after massage and bed treatment for a few weeks, the child is allowed to go about on a Taylor hip splint. In one instance the child is at present walking about with a perfect result, there being only an eighth of an inch shortening, which is due to a congenital malformation of the neck of the femur. This is the condition at the end of twenty months from the time of operation.

E. H. B.
H. L. B.

THE treatment has been wholly symptomatic. There is at the present time no known method of cure, and our efforts have been directed to alleviation and support. No form of internal medication has proved of special efficacy, though in certain cases cod-liver oil and other tonics may be of value. In the earlier stages of the disease, in which the diagnosis is often difficult, all methods tending to the strengthening and improvement of the general condition of the patient are advisable. Neither in this nor in the later stages of the disease have we found electricity of much value. We have not yet had the opportunity of testing the value of skilled massage applied regularly for considerable periods, but are now prepared to do so if suitable cases present themselves.

When the disease has reached the typical stage, it is sometimes advisable to consider the question of mechanical support; but the rule is, that the patient should be encouraged to get on without such help, and to gain exercise by the use of his limbs as long as possible. In the very late stages, the weakness or paresis of the limbs may become so great that no apparatus can be borne, and the patient becomes totally helpless.

W. N. B.

HARE-LIP.

COMPARATIVELY few cases of hare-lip have been treated at the hospital, as children under two are not as a rule admitted. In the twenty or more cases which have been treated, the operation has been done according to the ordinary methods described in the text-books of surgery, and usually with success. Instead of hare-lip pins for securing apposition, the practice has been in many cases to use buttons, one button being applied at each side of the line of sutures secured by catgut. The incision has been sewn with catgut, with the exception of two or three stout silk sutures. In three cases secondary operation was required, primary healing not taking place. In these cases the failure seemed to be due to the fact that the under side of the lip was not sufficiently sewn together, and the movement of the tongue of a restless child, forced the wound open after a few days. Since this experience the lips have been sewn both on the inside and the outside. Five cases of double hare-lip, with the projecting inter-maxillary bone, have been treated. In these cases the usual operation has been done with success, — namely, a V-shaped portion of the septum of the inter-maxillary bone is taken out, and the projecting bone is forced back into its normal position, the skin flaps are sutured, using the skin of the inter-maxillary bone with pared edges as a central point for the union of the flaps from both sides.

In some of these cases both sides have been operated upon at one sitting, and in others, at two sittings. They have all been successful. The dressing of the wound has been simple; the application of iodoform or aristol powder, and the daily painting of the wound with sterilized glycerine, and, in some instances, the application of a moist corrosive sublimate compress. The patients, as a rule, have been kept upon their back, lying upon a frame for a few days following the operation. In a few instances the wound has been infected by nasal discharge, and a second operation was necessary. To prevent this the insertion into the nostrils of iodoform wick is of advantage

E. H. B.

CLUB-HAND.

THERE have been five cases of this deformity treated in the hospital, two of which are specially worth reporting. In one case the infant was etherized, and the hand manually rectified and placed in an over corrected position, and held by a plaster-of-Paris bandage. The pain attending this was considerable for a few hours. The plaster-of-Paris bandage was renewed every few weeks, and the hand was held in its corrected position for a year or more, until it had grown into its normal position. A small tin splint was fitted to the internal surface of the forearm and hand in the latter part of the treatment, to serve as a substitute for the plaster-of-Paris bandage.

In the other case, where the right hand was flexed as well as inverted; the infant was etherized, an Esmarch bandage applied, and incisions were made on the palmar surface of two of the fingers; tenotomies were performed on the flexor tendons at about the level of the metacarpal articulations. It was also found necessary to perform tenotomy of the palmaris longus, and under antiseptic dressings, with the hand fixed in a wooden posterior splint, a satisfactory result was obtained.

The first of these cases has been under observation for three years, and the child uses the hand fairly well; the second case was under treatment eighteen months, when the result was satisfactory. The final result in both cases was very good.

H. L. B.

SEPARATION OF THE EPIPHYSIS OF THE HEAD OF THE FEMUR.

TWO cases of this rare deformity were brought to the hospital with a history of a fall followed by severe pain for three or four weeks, and lameness subsequent to this. In neither of the cases had the lesion been recognized. In one, there was shortening of an inch, in the other of half an inch. In one, there was marked inversion; in the other, there was but little.

In the first, a girl of twelve, osteotomy of the neck of the femur was done, and the limb brought into a normal position, and treated by traction for several weeks. The result was satisfactory; but the shortening of one inch was not entirely overcome. Perfect motion at the joint was re-established six months after the operation. The eversion of the foot, which constituted a disfiguring deformity, was entirely corrected.

In the second case, a boy of four, there was no noticeable eversion, and no operation was attempted.

E. H. B.

SARCOMA OF THE FEMUR.

BUT few cases of this disease have been treated at the hospital, and thus far amputation has not been done, as the cases were either too far advanced to offer hope, or the parents, in the earlier cases, refused operative interference. The important point in these cases is an early diagnosis, at a time when treatment may be of some avail, but unfortunately at this stage they present but little to differentiate them from cases of early joint disease. Thus far these cases have given a history of moderate pain, a slight limp, and some tenderness, usually appearing after an injury or some violence. Clinically, the condition presented is one of irritation of the joint, and the pain, although pronounced, is not a prominent feature. Enlargement appears early, and is a symptom of great importance, but is frequently difficult to distinguish from a deep abscess.

As an example, one case presented swelling with a sense of deep fluctation upon the outer part of the upper third of the thigh, and the symptoms resembled those of early hip disease. No enlargement of bone was apparent, and an incision was made to evacuate the contents of the swelling, when it was found that the upper third of the thigh was infiltrated with a growth, which proved to be a rapidly growing sarcoma. It was with difficulty that the hemorrhage was restrained, and the operation was abandoned for the time. Further treatment was refused by the parents.

E. G. B.

IMPERFORATE ANUS.

THERE have been four cases operated upon at the hospital. Of these, three were successful; one died. In one case a condition existed which was of peculiar interest. The child was four years of age,

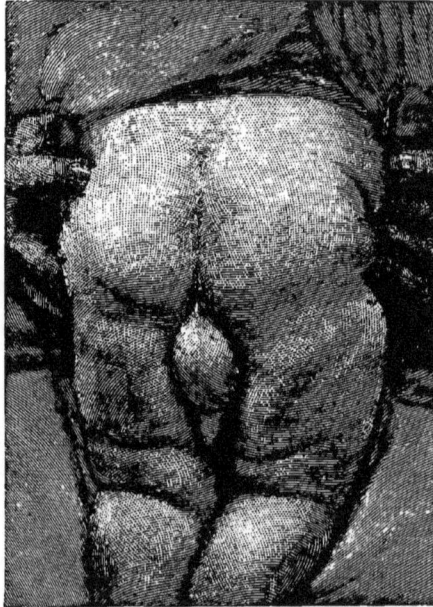

IMPERFORATE ANUS AFTER OPERATION.

and immediately after birth an incision had been made, and a trocar had been pushed blindly into the pelvis until meconium had been reached. This gave relief, but the opening was insufficient, and when the child presented itself at the hospital there was a constant dribbling away of faeces. The child used at times twenty napkins in the course of the day. Under ether it was found that the original opening with the trocar had been made outside the external sphincter. An opening was made in the centre of the external sphincter, the lower end of the bowel was found, brought down, and a successful result obtained.

In another case where there was only a dimple representing the anus, an excision was made of the coccyx, exposing the lower end of the rectum. This showed the obstruction in the rectum, which was two inches and a half from the anus; a careful dissection was then made up through the anus and through the diaphragm which obstructed the passage of the faeces; instant relief was obtained, and, at the end of two years, a permanent cure exists. The above cut shows the condition of the child at one year of age.

H. L. B.

AMPUTATION AT HIP JOINT.

AMPUTATION at the hip joint has been done a few times at the hospital. It has been proposed in a few other instances, but declined by the children's parents.

This procedure would undoubtedly save a number of cases which die from extensive pelvic disease, where proper treatment in the earlier stages has been neglected.

The operation at the hospital has always been done after a previous excision. The limb is deprived of blood by means of an elastic bandage, and a rubber tube is placed around the hip. A circular incision is made at the junction of the upper and middle thirds of the thigh, accompanied by a straight incision along the outer side of the upper third of the thigh. The flaps are easily separated, and the limb is removed with but little if any loss of blood.

In cases where this operation has been done, and extensive pelvic caries was found, the wound was packed with iodoform gauze, and left to granulate from the bottom.

In one case where complete cure took place, the patient, ten years later, showed a new formation of bone in the stump.

E. H. B.

PART III.

A STUDY OF TRACTION IN HIP DISEASE.

BY E. H. BRADFORD, M. D., AND R. W. LOVETT, M. D.

THE purpose of this paper is to present a study of traction as applied to hip disease; and the conclusions resulting from this study should be formulated as the answers to the following questions:

1. Does traction distract, — that is, does it draw the head of the femur away from the acetabulum?

2. If distraction does occur, under what circumstances does it take place, and what are its limiting conditions?

3. Can traction be employed with benefit as a means of treatment in hip disease?

The evidence presented will be

> *a.* Experimental,
> *b.* Pathological,
> *c.* Clinical.

a. 1. EXPERIMENTS IN TRACTION ON THE CADAVER.

Normal Joints. — In a number of dissecting-room adult cadavera the legs were slightly flexed and abducted, and a strong pull, varying from fifty to sixty pounds, was made upon the legs. It was found that the limb could be lengthened three-fourths of an inch. The cadavera were the ordinary subjects of the ana-

tomical room prepared for dissection, but not dissected.
They were in good condition. The same manipulation
was tried on a fresh cadaver where rigor mortis was
present, but no lengthening was obtained.

The cadaver of a negro aged forty years was subjected
to fifty-five kilogrammes of downward traction on the
right leg, which produced distraction of the joint surfaces. This was evident by an increase of distance
between the trochanters of eleven-twentieths of an inch.
On the left side, traction of twenty-five kilogrammes
produced a widening of nine-twentieths of an inch between the trochanters; with traction of sixty kilogrammes, fifteen-twentieths of an inch widening was
observed.

It was made manifest from these experiments, and from
two partially dissected specimens, that a certain amount
of separation of the head of the femur from the acetabulum was possible, if traction of fifteen pounds was
made with the limb in the position of slight flexion
and abduction, but where the limb was kept straight no
amount of force caused distraction in adults. This was
true in both dissected specimens, in one of which the
skin and muscles had been removed.

The hip of a full-termed fœtus was prepared in such a
way that the skin was removed so as to expose the
muscles around the hip. It was found that under a
slight amount of traction distraction was possible. This
was not only visible to the eye, but it was also demon-
strable on a specimen on which the skin was removed
without disturbing the ligaments or muscles. A needle
was inserted in the head of the femur, and another in the
ilium slightly above the acetabulum, a slight amount of
force separating the two needles.

An adult dissecting-room specimen was taken, the femur amputated below the trochanter, and the pelvis fixed. The skin was not removed, and a traction force was applied. Needles were inserted into the femur and into the ilium, the skin and muscles being incised in such a way that the traction force would not disturb their relative position. Traction of a hundred pounds was applied, and it was found that the needles were separated an eighth of an inch. After the specimen had been soaked in weak alcohol for some time, distraction of an eighth of an inch was easily effected by a pull of five pounds.

It was clearly shown that traction distracted in all cases in children, dissected or undissected, and in all specimens of infants, and that the checks to distraction in the cadavera of adults lay in the resistance, first, of the capsular ligament, especially of the anterior bands of the ilio-femoral ligament; second, in the resistance of the cotyloid ligament, and to a slight degree in atmospheric pressure. In children the lower edge of the acetabulum presents no resistance to traction in the line of the axis of the body. In adults this presents a resistance, but if the limb is abducted the resistance is avoided. Both in children and in adults, if the femur is extended to its utmost limit, the anterior bands of the ilio-femoral ligament lying on the front of the capsule, prevent all distraction by any force which it is feasible to apply. If the capsule and cotyloid ligaments are disorganized, distraction is easy.

Diseased Joints. — In a specimen of a case of hip disease of six months' duration, where death took place from scarlet fever, it was found that distraction was easily made by the slightest traction. In this specimen

the cotyloid ligament was disorganized, but the strong
ligamentous fibres of the capsular ligament alone served
as a check to separation of
more than half an inch on
traction. But within that
limit even the weight of
the pendent fragment of the
femur distracted, as is seen
in the accompanying illustration.

In experiments upon the
cadaver, however, it must be
remembered that the element
of muscular contraction is
absent. This is, next to
the anatomical condition, the
greatest obstacle to distraction; and what might be perfectly true of the cadaver
might fail in its application
to the living subject, whether
having a diseased or a healthy
hip joint. Consequently, experiments upon the cadaver
must be confirmed by experi-
ments upon the living, if they are to be of value.

SPECIMEN SHOWING DISTRAC-
TION OF THE HIP.

[*From the N. Y. Med. Journal.*]

a. 2. EXPERIMENTS UPON THE LIVING SUBJECT.

The experiments here reported were made at the
hospital.

The method of experiment was as follows: The patient
was placed upon a hard table with the head against the
wall, and perineal straps upon each side were secured to

the head of the table by stout webbing. In some instances shoulder straps of a similar character were also added. This was for the purpose of preventing the child from slipping on the table as far as possible. All measurements were taken from the wall. Measurements at various points were taken by different observers. The anterior superior spine was marked with a hair line in ink on both sides, and in some of the experiments the great trochanter was marked as well. A mark was also made at the site of the external malleolus. A tape was carried from the wall touching these marks on the side experimented upon, and on the other side it was carried to the anterior superior spine to show any tilting of the pelvis which might occur. Traction on the leg was made by means of webbing straps fastened to a lacing which did not go below the knee. Traction, therefore, was made wholly upon the thigh. Traction was made by means of a spring balance fastened to the webbing straps below the foot. In each experiment traction was first made of ten pounds; then of twenty pounds. To prevent any error caused by the slipping of the skin around the sole of the foot, a plaster-of-Paris bandage or a stout cotton bandage was applied from the toes to the knee, and upon this bandage the site of the external malleolus was marked. The heel was made to slide upon a glass plate to avoid friction. In making the experiments any case where the heel left the plate during the traction was thrown out as inaccurate. The experiment was made as follows :

An observer was detailed to watch the mark made over the anterior superior spine ; another observer was detailed to notice the mark at the external malleolus ; a third noted the anterior superior spine on the well side, and in some of the earlier experiments, to check the correctness of the method, independent observers were placed either at the knee or at the great trochanter. In most instances three observers were employed, one at the anterior superior spine, one at the external malleolus on the diseased side, and the other at the anterior superior spine on the well side. (See illustration on page 303.)

The patient was placed upon the table as prepared. Each observer read the position that the line, marked with ink upon the part of the patient he was to watch, measured on the tape. Traction of ten pounds was made. Each observer noted the position under the new conditions, and they were put down by the recorder. Traction of twenty pounds was made, and each observer noted the position of the line on the tape. These were also noted by the recorder. In every experiment, unless otherwise stated, the experiment was immediately verified with the observers changed. The method of observation, in short, was to measure the distance of the external malleolus from the wall; knowing the distance of the anterior superior spine, to make traction upon the leg, see how much the external malleolus had descended; then, noting how much the anterior superior spine had been pulled down, to find the amount of separation between the external malleolus and the anterior superior spine, this giving the amount of distraction of the hip joint surfaces. The method of these experiments has been related in detail because upon its accuracy the value of these experiments depends.

Various sources of error were eliminated. The fact that traction was made upon the thigh alone eliminates any source of error from stretching of the knee-joint ligaments.

An error due to the stretching of the skin may be dis-

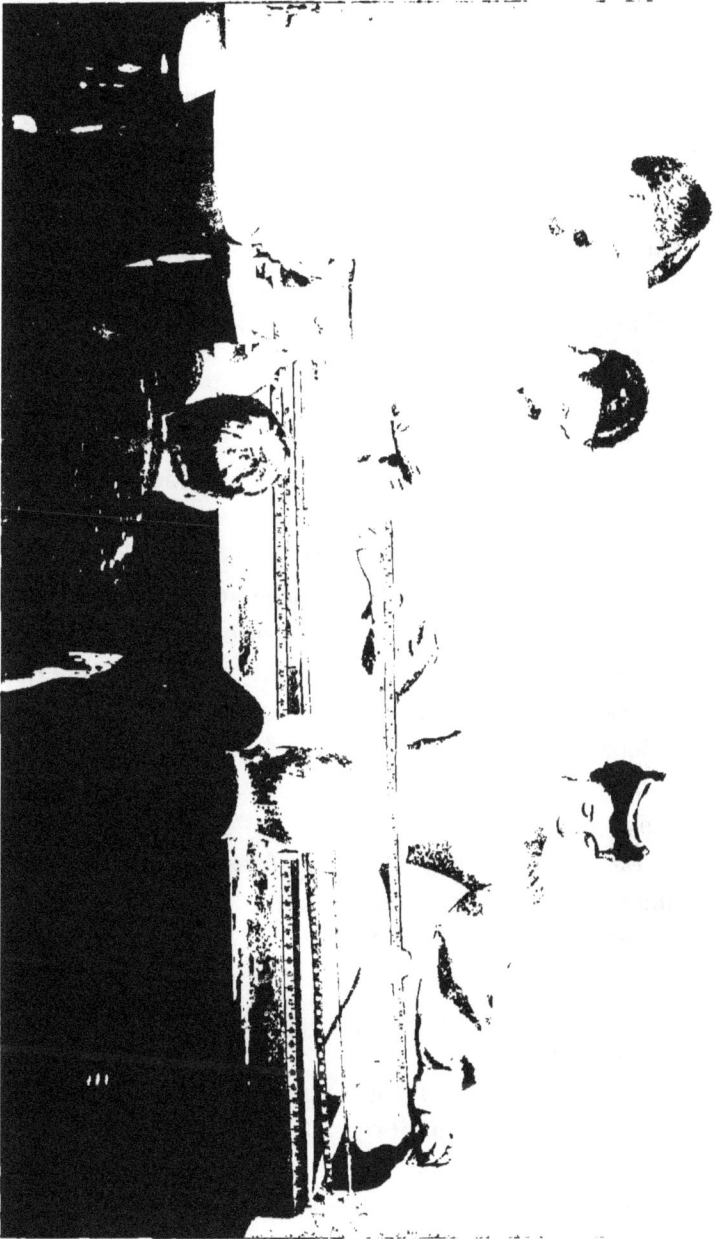

METHOD OF MAKING THE TRACTION EXPERIMENTS.

[*From the N. Y. Med. Journal.*]

regarded in these observations. The skin of the thigh is pulled down, but the skin of the leg is not pulled upon. Consequently, any such stretching would tend to show less lengthening than really occurred.

Observations upon Healthy Joints. — The first experiment, which is of special interest, is not in the table. A girl of seven, with dorso-lumbar Pott's disease, had an abscess which pointed at the outer side of the thigh. This was opened by an incision of three inches, exposing the trochanter. The hip joint was healthy. Some days after operation the girl was laid upon a table, secured in place, and an upright was erected upon the table with the needle pointing at the marked spot on the exposed trochanter. Ten pounds of traction produced no measurable effect; traction of twenty pounds produced lengthening of a quarter of an inch, as seen by the mark on the trochanter as compared with the fixed point adjacent, — *i. e.*, the needle. If traction of twenty pounds was made, the head of the trochanter could be seen to descend; if traction was suddenly relaxed, the head of the femur could be seen to move upward.

Observations upon Diseased Joints. — In these experiments traction was made in the line of the body, and, unless otherwise stated, the amount of malposition present was not enough to be noted.

As evidence of accuracy of these measurements it is to be remembered:

1. At the time of the experiment the observers were entirely ignorant of its result.

2. The error caused by the slipping of the skin tends to diminish the amount of distraction as shown by these experiments.

TRACTION IN HEALTH.

Case No.	Sex.	Age.	Condition.	Traction in pounds.	Result in inches.	
1	Male.	6 years.	Hip disease on other side. Healthy hip examined.	10 20	⅛ lengthening. "	Verified by change of observers on repeated experiment.
2	Male.	7 years.	Hip disease on other side. Healthy hip examined. Hip disease on other side.	20 10 20 10	" " " "	Verified by change of observers on repeated experiment. Not verified.
3	Female.	7 years.	Healthy hip examined. Hip disease on other side.	20 10	⅛ shortening. ⅛ lengthening.	Verified by change of observers.
4	Male.	7 years.	Healthy.	20 10	" "	Verified by change of observers.
5	Male.	10 years.	Healthy hip examined. Hip disease on other side.	20 10 20	No change. No change. ⅛ lengthening.	Verified by change of observers.
6	Male.	12 years.	Two observations on healthy hip: First experiment . . Second experiment . .	10 20 10 20	No change. No change. ⅛ lengthening. No change.	Verified by change of observers. Verified by change of observers.
7	Male.	16 years.	Healthy hip examined. Hip disease on other side: First experiment . . Second experiment . .	10 20 10 20	No change. ⅛ shortening. No change. ⅛ shortening.	Verified by change of observers. Verified by change of observers.

3. The experiments agree with each other and with those of other observers.

The experiments in general need no comment, except that it is interesting to note that in Experiment 8 the child had never had traction applied before, and in that case the largest amount of distraction occurred. That is to say, it seemed as if in the other cases where traction treatment had been used, a certain amount of previous stretching of the muscles might have existed.

The conclusions which can be drawn from this table seem to be the following: That traction of ten pounds in children before puberty as a rule produces lengthening of the leg in hip disease, and that this lengthening is due to separation of the joint surfaces; that the amount of this separation varies in different instances, being in general less in older children than in young ones, and also varying in individual cases under apparently the same conditions, perhaps on account of some anatomical peculiarity; that twenty pounds traction, as a rule, produces more separation than ten pounds.

It is probable that late in hip disease, where cicatrization of the capsular tissue may have taken place, distraction is not as readily made.

b. Pathological Evidence.

An opinion of the value of traction should be based not upon experiments alone, but also upon pathological evidence.

The pathological process characteristic of hip disease is illustrated in a number of pathological specimens seen in the Warren Museum, which are found to resemble each other in presenting the characteristic changes, varying

TRACTION IN DISEASE.

Case No.	Sex	Age	Length of Disease	Character of Disease	Amount of traction in pounds.	Result in inches.	
1	Male.	5 years.	7 months.	Acute.	10 / 20	No change. / lengthening.	Verified.
2	Female.	5 years.	3 months.	Acute and sensitive.	10 / 20	" / lengthening.	Not verified on account of pain.
3	Male.	4½ years,	1 year.	Quiescent; fifteen degrees of motion.	20 / 10	No change. / lengthening.	Verified with different observers.
4	Female.	6 years.	3 years; sinuses.	Acute; no malposition; few degrees of motion.	20 / 10	"	Not verified.
5	Male.	6 years.	2¼ years	Convalescent; old abscesses.	20 / 10	"	Verified.
6	Male.	7 years.	3 years.	Very sensitive; abscess, spasm slightly abducted.	10 / 20	"	Verified.
7	Male.	7 years.	3 months.	Acute; some motion.	20 / 10	"	Verified.
8	Female.	8 years.	1 year.	Acute and spasm; not very painful.	10 / 20	"	Verified. Never had traction applied before.
9	Male.	10 years.	3 years.	Moderately sensitive; very little motion.	10 / 20	No change. / lengthening.	Verified.
10	Male.	10 years.	3 years.	Not sensitive; forty-five degrees of motion.	20 / 10	"	Verified.
11	Male.	12¾ years.	3½ years.	Convalescent; good motion	10 / 20	"	Verified.
12	Male.	16 years.	Indefinite; over a year.	Forty-five degrees of motion.	10 / 20	No change. "	Verified.

only in the extent of the destructive osteitis. (See illustration.)

Erosion of the Upper Part of the Acetabulum.
[From the N. Y. Med. Journal.]

The change from carious destruction is most marked in the upper portion of the acetabulum, and in the lower portion of the acetabulum there is evidence of repair in some of the specimens.

From the specimens examined it is clear that in hip disease the head of the femur is crowded against the acetabulum in a direction upward and backward, and that the process of repair is more advanced where the pressure is removed.

Physiologists estimate the force of a muscle fully contracted at from six to ten kilogrammes to every square centimetre of muscular surface on cross section. In an adult, at the hip joint, the muscles connecting the femur with the ilium may represent from ten to fifteen square centimetres, and although these muscles are rarely contracted to their full extent, it is evident that the amount of force when slightly contracted is by no means inconsiderable; and during an acute spasm, when the muscles are firmly contracted, the pressure driving the head of the femur upon the acetabulum must be very great even in a child. It is well known that the muscular spasm at its acute stage is both a tonic spasm and also an acutely exaggerated spasm on any jar or violence to the hip, or even on the apprehension of any jar or violence. This spasmodic stage subsides after a while if the hip is kept absolutely free from motion, but it is a matter of experience that this spasm may persist for months, reappearing upon locomotion until the morbid process is entirely corrected and the inflamed bone is solid.

The effects of traction, when thoroughly carried out, can be seen in the accompanying specimens.

The first is that of a boy of nine, who was attacked with hip disease of an acute form six years before. He was treated with traction efficiently for a long time, first with recumbent fixation, later with an ambulatory traction splint and crutches, and afterward by a protection splint. An abscess developed in the early stages, was incised, and subsequently healed entirely. The boy

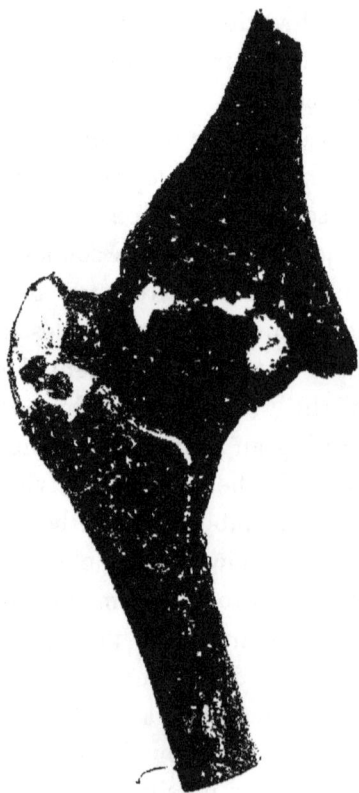

SPECIMEN OF HIP DISEASE TREATED BY TRACTION.
[*From the N. Y. Med. Journal.*]

recovered completely after a number of years from his hip disease, having, however, a limb which was slightly shorter (an inch and a half) than the other with limited motion. The position was good, and the leg was thoroughly useful and remained so two years after the discontinuance of all treatment, the boy being as active

as any boy at this time. He was, however, subsequently
seized with tubercular meningitis, and died. At the
autopsy complete cure of the hip disease was found, and
this specimen also shows that there has been no widening of the acetabulum, and but little alteration in the
shape either of the acetabulum or head of the femur.
(See illustration, page 310.)

A comparison of this specimen with those of severe
hip disease where traction was not used speaks most
emphatically for the thorough use of the method.

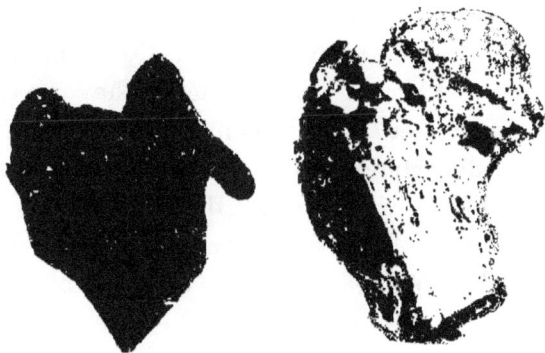

SPECIMEN ON THE LEFT FROM A CASE NOT TREATED BY TRACTION; THAT
ON THE RIGHT FROM ONE TREATED BY TRACTION.

[*From the N. Y. Med. Journal.*]

The specimen on the right is the head and neck of the
femur where excision was done after two or three years
of efficient treatment by traction, but the reparative process was not sufficient in this case to establish a cure;
the patient's general condition failed, and excision was
done. It is to be noticed that there is very little alteration in the shape of the head of the excised femur.
This, compared with the accompanying specimen of an
excision of a patient with hip disease of similar severity

and duration where no traction had been applied, would appear fairly to show the effect of traction in saving the head of the femur from destruction.

c. CLINICAL EVIDENCE.

The cases reported below were taken from the records of the hospital, and the patients have been under the care of the surgeons of the hospital, all, however, carrying out treatment by more or less efficient traction during the requisite stages. They represent cases where, from the history of the results, there could be no doubt as to the existence of well-marked disease in the joint, and are selected because of this fact. They are all cases which were treated at their homes under the direction of the out department, after their discharge from the wards where they were treated during the acute stages when necessary. They do not represent the best results which can be obtained under the direction of a trained nurse or an intelligent mother. They are hospital cases treated in a routine way. They are intended to illustrate the fact that in cases thoroughly and properly treated by traction subluxation can be prevented; and that in cases of the severer types, if treated early, some motion of the hip joint can be preserved.

In the cases here reported, the diagnosis of hip disease was certain. The record of motion is without doubt in the cases where it is recorded, as it was made with particular care, and all cases were rejected where there was any doubt. The motion was tested by placing the patient on the back, with one hand upon the pelvis, the other manipulating the thigh. The cases had all been under observation for a long period.

The cases may be grouped : First, as those of hip disease of a severe type, as proved by the development of abscess or the arrest of growth ; second, cases without abscess, but with persistent spasm, limitation of motion, and deformity, and a long period of pain and sensitiveness; third, the lighter form of disease treated before the severe symptoms had developed. These cases may be regarded as representative ones seen at the hospital where continued treatment was carefully carried out.

1. *Cases of Severer Type.*

CASE I. — Annie F. entered the out department of the hospital in February, 1888, being at that time fifteen years old. The disease had been in progress for two years, one of which had been spent in bed. Pain had been severe, and night cries frequent. An abscess had formed, and the joint was flexed and fixed. Traction treatment was begun and continued for two years with traction splint and crutches. A protection splint was worn for four years more.

Present Condition. — Twenty-one years old ; strong, healthy woman ; weight, one hundred and twenty-one pounds. The sinus had been healed three years. There is motion in flexion of ten degrees at the hip joint. There is no motion in other directions. Patient walks well. There is a three-inch shortening, but the trochanter is not above Nélaton's line. There is no deformity.

CASE II. — Nellie M. entered the out department of the hospital in September, 1884, when eleven years of age. The disease had lasted for three years. There had been much pain, and the patient had been treated by high shoe and crutches. Abscesses had been present, and a sinus remained. Persistent muscular spasm and pain. Traction treatment was carried out, and a traction splint worn for three years and a half; after this a protection appliance was worn and is still worn, although no symptoms have been present for a long time.

Present Condition. — Twenty-one years of age, strong and healthy. Walks firmly without splint, but with a limp. The trochanter is below Nélaton's line. There is shortening of two inches from difference in growth. Motion of the joint limited except in flexion.

CASE III. — George K. entered the out department of the hospital in March, 1887, when fifteen years and a half old. The disease had existed for four months. Traction splint was applied. The hip became sensitive, and an abscess appeared the following year. Muscular spasm lasted for two years and a half. Traction was continued for three years, and a protection splint worn four years longer.

Present Condition. — Twenty-two years old ; healthy, strong man, walking without a splint. There is an inch and a half shortening of the leg, but no subluxation, the trochanter being below Nélaton's line. The position of the leg is normal. There is no motion.

CASE IV. — Hattie H. came to the out department of the hospital in March, 1886, when five years old. Disease was of six months' duration. The leg was flexed to an angle of forty-five degrees. There was much pain and sensitiveness. The muscular spasm continued for nearly two years, and an abscess followed. Traction treatment was carried out for two years, a traction splint being worn a good portion of the time. A protection splint was used for three years more.

Present Condition. — At the age of thirteen the child is strong and well. The trochanter is below Nélaton's line. There is shortening of half an inch. Flexion of ninety degrees is possible. Walks without a limp. There is no deformity.

2. *Cases of the Second Class.*

CASE V. — Sophie R. entered the out department of the hospital in January, 1886, when six years of age. The disease had lasted for nine months, and the hip was fixed. There was pain, and the spasm lasted for two years. Treatment by traction was carried out for three years and by protection for two years more. No abscess occurred.

Present Condition. — January, 1893, there was half an inch shortening. Flexion was possible to a right angle. Rotation and abduction limited.

CASE VI. — Clara L. came to the out department of the hospital in March, 1888, when seven years old. The disease had existed for two years. At the time when first seen at the hospital there was a distortion; the leg was abducted, fixed, and very sensitive, with persistent pain and sensitiveness. The muscular spasm lasted for three years. Bed treatment and admission to the hospital were required for pain and sensitiveness. No abscess developed. Treatment by traction was continued for three years and a half. A traction splint was worn for three years, protection splint for four years and a half afterward.

Present Condition. — The patient is thirteen years old, strong and well; slight motion at the hip joint. There is no deformity except slight permanent flexion. The diseased limb is two inches shorter than the other; the trochanter, however, is below Nélaton's line.

CASE VII. — Robert H. was brought to the hospital in March, 1888, when four years old. The disease had lasted about two months. There was much muscular spasm at the hip, with marked pain, which persisted for some time, with swelling about the hip. Bed treatment was carried out for a month. The muscular spasm improved after six months, but remained for two years. Traction treatment was applied during all that time, and a traction splint worn while the patient was up. A protection splint was worn for two years more.

Present Condition. — At the age of ten the patient walks without a limp. There is shortening of half an inch in the affected limb, but no deformity. Motion is possible to ninety degrees in flexion; rotation is limited.

CASE VIII. — Esther M. came to the out department of the hospital in 1888, when eight years old. Disease had lasted for six months. The hip flexed and adducted. Pain was severe. No motion at the hip joint was possible. Pain and sensitiveness were marked, and bed treatment necessary. Treatment by traction was carried out for three years, and protection for three years more. Protection splint is still worn as a precaution.

Present Condition. — The patient is fourteen years of age, strong and well, and can walk without a splint. Forty-five degrees of motion is possible in the direction of flexion. There is an inch and a half of shortening, but the trochanter is not above Nélaton's line. There is no deformity.

CASE IX. — Lizzie C., brought to the out department of the hospital in May, 1886, when eight years old. Disease had lasted six months. Leg was fixed and abducted, and there was no motion. Muscular spasm continued for five years. There was no abscess, but patient required entrance to the hospital and bed treatment several times. Traction treatment by means of weight and pulley and traction splint continued for six years; protection for two years more.

Present Condition. — Sixteen years of age, strong and healthy girl, with shortening of half an inch. Ten degrees of motion possible at the hip joint. There is no malformation nor deformity. Can walk without pain, but at times wears the protection splint.

CASE X. —Anastasia H. entered the hospital in 1886, when five years old. Disease had been in progress for several months. Night cries had been noticed for three months. Admission to hospital for bed treatment. Patient remained in hospital three months. There was no abscess. Spasm continued for two years. There was pain and persistent adduction. Traction treatment carried out for two years and a half; protection for a year and a half longer.

Present Condition. — Thirteen years old; girl is strong and well, walks without a splint and with no perceptible limp. There is an inch shortening, but no deformity. Flexion of ninety degrees possible, but limitation in other motions.

CASE XI. —Nellie M. C. entered the out department of the hospital in April, 1886, when five years old. Disease had lasted six months. Hip was flexed and fixed at an angle of forty-five degrees and very sensitive. Spasm remained for two years. Traction was carried out for two years and a half, and protection for five years longer.

Present Condition. — Child thirteen years of age, strong and well. There is a permanent flexion of ten degrees, but no deformity. There is a shortening of an inch, and the child walks with a limp, but needs no apparatus.

3. *Cases treated at an Early Stage.*

CASE XII. — James G. entered the out department of the hospital in April, 1890, with a history of pain in the knee at night for several weeks. Pain continued for some time. Limitation of motion. There was, however, but little muscular spasm. A traction splint was applied and worn continuously for two years.

In August, 1892, a protection splint was applied, and has been worn since that date.

Present Condition. — The position of the leg at present is normal. There is no shortening. Motion beyond ninety degrees. There is no muscular spasm.

The diagnosis in this case was based upon the pain which persisted, the limitation of motion, and the length of time which the muscular spasm lasted.

CASE XIII. — Eva C. The patient entered the out department of the hospital November, 1891. There was severe pain, with night cries, muscular spasm, and deformity, and these symptoms had persisted for several weeks. The patient entered the hospital and remained in bed with traction treatment for six weeks. A traction splint was worn for a year, and then removed by the parents, the child being considered by them in perfect health. The child was allowed to use the leg freely, and a relapse occurred after six months, with pain, night cries, spasm, and deformity. Traction treatment was renewed after a preliminary bed treatment with fixation and traction.

Present Condition. — At the present time, three years and a half after commencement of treatment, there is slight permanent flexion and free motion of twenty degrees. There is no subluxation and no shortening. Patient still wears a traction apparatus.

This case is reported as indicating a lack of perfect result. Treatment was discontinued by parents for several months, and a relapse occurred.

The case is still under observation ; but the ultimate result, which could in all probability have been without a limp, will be with a slight limp.

The questions formulated at the beginning of this paper may, from the evidence presented, be answered as follows :

1. Traction properly applied can and does draw apart the surfaces of the hip joint, in the cadaver and in the living subject.

2. A greater amount of traction force must be applied

than has commonly been used in hip disease. Distraction is less likely to occur in the adult than in the child,
and more likely to occur in diseased or disorganized
joints than in healthy ones. Distraction is more easy in
a flexed and abducted position of the limb than in a position of full extension.

3. Pathological evidence demonstrates that the upper
edge of the acetabulum and the head of the femur are
eroded in hip disease where traction is not used. In the
specimen shown, where traction was employed, this had
not occurred.

Clinical evidence shows the character of the results in
routine hospital cases after the use of traction in the
absence of subluxation of the femur, and in the preserva-
tion of motion.

Upon pathological grounds, and in view of the demon-
strable results of cases treated by traction, it may be
stated that traction may be employed with benefit in the
acute stage of all cases of hip disease.

SUGGESTIONS AS TO THE CAUSE OF BOW-LEGS AND KNOCK-KNEES.

BY HERBERT L. BURRELL, M. D.

FOR the past fifteen years the writer has been interested in bow-legs and knock-knees as occurring in children. There are several well recognized facts regarding this condition. They are: —

1. That a large number of children between two and five years of age are slightly bow-legged or knock-kneed.

2. That very few adults are bow-legged or knock-kneed.

3. That the indication for operation in these conditions has not been clearly defined, and that to do an osteoclasis or an osteotomy upon a child because its legs are bowed, and it has passed three years of age, are unsatisfactory reasons for operating.

4. That in a given case of bow-legs or knock-knees in a child between three and five years of age, it is largely a question of personal opinion whether the child's legs will correct themselves or not.

Prior to the introduction of osteotomy and osteoclasis, crooked legs in adults were uncommon, and it has been a question in the mind of the writer whether unnecessary operations are not performed for these conditions. To determine the causation and life history of bow-legs and knock-knees, it was decided to take tracings of all cases as they presented themselves, and some hundreds of observations have been made, extending over seven years. Of this number a small proportion have corrected themselves, and a large number have been operated upon. There are very few general practitioners of twenty or

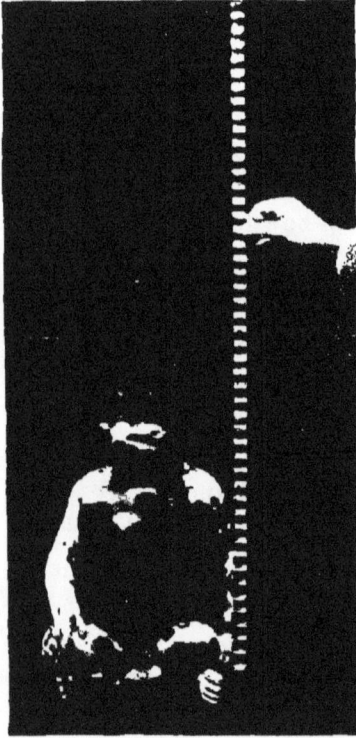

Child sitting in Turk fashion, producing at junction of lower and middle thirds of legs anterior and lateral bowing.

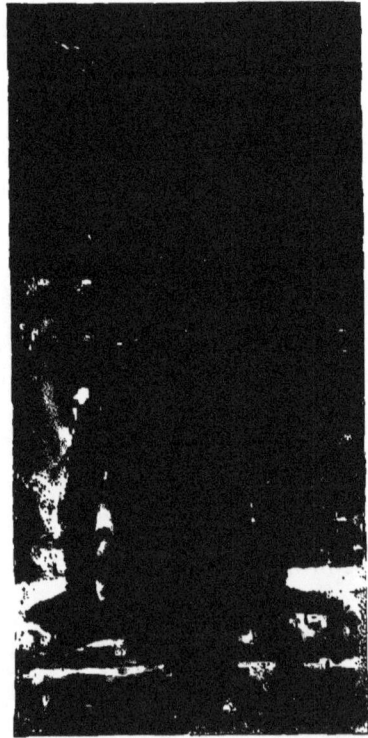

Child with bow-legs, sitting in its ordinary position, showing the fitting of one leg to the other.

more years' experience who cannot relate numerous cases of the self-correction of bow-legs in young children.

A study of the types of the deformity as they existed was made from photographs and tracings. It suggested itself to the writer that as these deformities occur between the first and the fifth year most commonly, and rarely occur after the tenth year, there must be conditions at that period of life which would account for the deformity.

Attention was paid to the attitude of infants and children, and inquiries made of the parents regarding the habits of the patients in sitting, creeping, and kneeling. It became evident that in posture an important factor existed in determining the kind of deformity. It was found that an infant of twelve months who is improperly nourished, its bones being soft, and who is allowed to sit with its legs crossed beneath it in Turk fashion, may have a deformity created in the tibia, consisting of an anterior and lateral bowing at the junction of the lower and middle thirds of the leg (see cut on page 320). The crossed legs often fit into one another as accurately as if they had been moulded one to the other, as in fact they have been (see cut on page 320).

Again, it is found that an infant or child who is allowed to sit in a low chair with its feet resting upon the floor, or a child seated in a " high chair" with its legs and feet resting upon a high cross-piece, the swaying to and fro of the child's body with the feet applied to the floor, or the high cross-piece, often produces an antero-posterior bowing at the juncture of the lower and middle thirds of the leg.

Again, when the child sits constantly in too high a chair, with its feet not touching the floor, an anterior bowing may take place at the juncture of the lower and middle thirds of the femur.

Attitude in children is often the cause of knock-knee. A child may acquire the habit of standing upon one foot and allowing the inner side of the sole of the other foot to rest upon the ground. This produces traction upon the internal lateral ligament of the knee joint in the deforming leg, and either produces relaxation of the internal lateral ligament or extra pressure upon the external condyle of the femur, thus relieving the pressure upon the internal condyle, which results in an unnatural growth

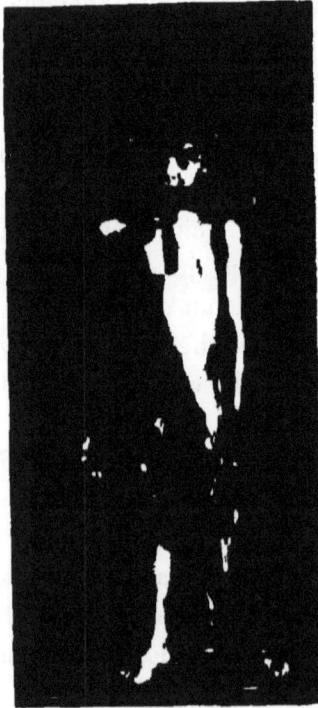

STANDING "AT EASE."   STANDING "AT ATTENTION."

at that point. The boy figured above shows the effect of attitude in forming knock-knee. In the first he is standing "at ease," as he ordinarily does; in the second he is standing "at attention," and has completely obliterated the beginning deformity.

It is obvious that as soon as the tendency to deformity is established, it goes on with increasing rapidity;

for while standing in a faulty position the weight of the body adds a constantly acting force.

It has been noticed in some children, two to three years of age, with soft bones and lateral bow-legs, that where they also had a protuberant abdomen, there was frequently an anterior bowing at the juncture of the lower and middle thirds of the femur or tibia; and cases have been watched where the abdomen has become prominent, until finally the weight of the body was thrown so far forward that the pelvis was projected backward, and there occurred a bowing of the femur and tibia at the juncture of the lower and middle thirds.

It was then thought that a cause of these deformities might be found in the distribution of weight upon the lower extremities. Bearing upon this causation are several interesting anatomical facts. It was found that the "splay" of the pelvis in infants is very different from that in adults. McAllister states that the "splay" of the pelvis in infants is 137°, in children 145°, in male adults 153°, and in female adults 155°. It was then remembered, and found statistically true, that bow-legs in girls correct themselves more surely than in boys. These facts suggest that in the relatively rapid change that occurs in the angle of juncture of the femur and pelvis (it being 137° in infants, and in children 145°), we might have a factor in determining these deformities. It was then remembered that the centre of the body changes as growth occurs, and through the kindness of Professor Dwight I was able to find the references to this anatomical fact. The following figure (page 324) taken from "Anatomie der äusseren Formen des menschlichen Körpers, von Dr. Carl Langer," shows schematically the variation. It will be noticed that the greatest changes occur between one and five years, the centre dropping one half the distance from the umbilicus to the pelvis, and as this is the period of life when bow-legs and knock-knees occur, this changing of the centre of the body seems to be an important factor in causation.

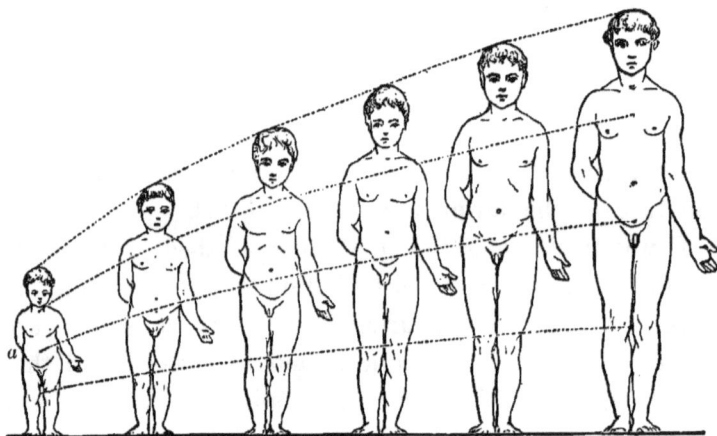

CENTRE OF BODY AT DIFFERENT AGES.
a is the line showing the changing centre of weight.

The effect of muscles in producing curves must be very great. A curve once started, or the normal one at birth persisting, it can be easily understood may increase alone by muscular action. Undoubtedly in many instances in the beginning of a curve, the normal balance between the strength of the muscle and the resistance of the bone is lost. In walking the muscles are developed, the group of muscles on the outer surface of the lower extremity are less developed in infants than the inner group of muscles; and as a child begins to walk and the muscles develop, the equilibrium between these two groups of muscles may be disturbed, and the stronger, inner muscles, the adductors, may start an outward curve.

The subject of muscular development of the legs is a very interesting one, but its effect has not been thoroughly studied.

In one family where all the children were bow-legged, it is interesting to note that the girls' legs straightened, while the boys' remained crooked into adult life.

The various supplementary or additional curves which occur in individual cases may be accounted for by habits of attitude in sitting, creeping, kneeling, and walking, or other factors besides superincumbent weight and muscular action.

It has been found that rickets is not present in all bow-legs, nor was a previous disease a constant factor.

A number of Italian children have been seen whose legs were straight before immigration, who developed marked deformity after coming to this country.

To consider this problem as a whole, the cause of bow-legs is complex, and among the many factors entering into their causation the following may be noted : —

1. The centre of the body changing rapidly between two and five years of age.

2. Change in distribution of weight through the pelvis and femora, especially marked between two and five years of age.

3. Muscular action.

4. Attitude in standing, walking, or creeping.

5. Malnutrition.

The result of treatment in a given case is largely dependent upon the length of time the bow-legs has existed. If the deformity has existed over a year, less hope of success can be had than after it has existed under a year. An operation should be considered if the bones are hard and unyielding, the child over three years of age, the curve a general one, and there cannot be found a faulty distribution of weight as shown by attitude. This opinion is formed because it has been found that the majority of patients whose legs have straightened themselves were between one and three years of age, they presented signs of general rickets, their bones were yielding, and their deformity was great or small, usually small.

The treatment under three years of age may be summed up as follows :

1. If the deformity is increasing and the bones are still soft, either absolute rest with fixation of the legs by plaster-of-Paris bandages in a partially corrected position; binding the legs together at night; or braces.

2. If the bones are hard and unyielding and there is no sign of improvement, operation.

The consideration of this subject by the writer is incomplete, and is purely suggestive. The tracings that have been taken of bow-legs and knock-knees in the past he finds are useless, on account of the changing weight centre and the distribution of weight. Hereafter a tracing of the whole child will be taken, the centre of the body and weight of the child will be recorded, to be used in determining the cause of a given deformity.

In this subject there is an important and interesting field open for research, and the writer hopes that in the future he may be able to work out the problem.

TENOTOMY IN SPASTIC PARALYSIS.

BY WILLIAM N. BULLARD, M.D.

IN the first half of the present century the operation of
tenotomy received a strong impetus and was brought
forcibly to the attention of all practitioners who devoted
themselves to the treatment of deformities of the limbs.
This was largely due to the advocacy of the subcutaneous method by Stromeyer, and to the writings of Dief-
fenbach, Little, and Bonnet. Dr. Henry J. Bigelow, in
the Boylston Prize Dissertation for 1844, gives a careful statement of the more advanced views on tenotomy
at that time. It was then generally advocated in the
more severe and protracted forms of club foot, whether
congenital or acquired; but no definite distinction appears to have been made as to its value in paralytic
and spastic contractions. It was not until some dec-
ades later that we find the attention of surgeons
and orthopedists drawn to this distinction, not indeed
until the general advance of neurology, with the discovery of the significance of the condition of the knee-
jerks and its interpretation of the pathological causes
of the various forms of paralysis, rendered it no longer
possible that this distinction should be overlooked.

Since 1870, however, the general professional opinion
among orthopedists and surgeons has, with due regard
to the distinction between these forms of contracture,
been, in general, that while tenotomy was of great
value, and its employment advisable and necessary in

many cases of paralytic contracture (congenital varus,
club foot, &c.), on the other hand in cases of spastic
paralysis due, as we now recognize, largely to intracra-
nial affections, tenotomy was a somewhat hazardous
procedure, offering little or no hope of permanent
improvement. It was generally held that in those
cases where tendons were severed, whether subcutaneously or by open incision, a recontraction was quite
certain to occur, and that rarely did any permanent
advantage to the patient result.

It was while the general professional opinion was of
this character, that Bradford, in 1885, published an article
on tenotomy in spastic cases, in the Boston Medical and
Surgical Journal, advocating this operation in certain
cases of this character, and announcing the results which
he had obtained. Some of the cases on which that article
was based were operated upon in the Children's Hospital,
and will be enumerated in our text. Since the publication of Bradford's article, a number of tenotomies for spastic contractures have been performed here, and it is our
purpose in this article to determine, so far as may be, the
results attained.

It is scarcely necessary to point out the close relation
between neurology and surgery in cases of this character,
nor the important bearing which the advisability or non-
advisability of operative procedure in this class of cases
has for every neurologist. There is a form of spastic
paralysis which is practically incurable by any known
means short of surgical operation, and the question to
be determined is whether these patients shall be considered incurable, or whether tenotomy and other surgical procedures offer a sufficient prospect of alleviation
or cure to render their recommendation justifiable.

In considering the results of tenotomy in cases of permanent spastic contraction, we must distinguish three different questions, — the dangers of the operative procedure itself, and its immediate surgical aspect; the probability of recontraction of the tendons with resultant recurrence of the old condition; the actual permanent condition of the extremity after operation, and whether this condition is more beneficial to the patient than that previously existing.

Spastic paralysis consists essentially in two conditions, — one, that of tonic spasm or contracture with its accompaniments ; the other, that of paresis or partial paralysis. Tenotomy is addressed solely to the remedying of the first condition, — the contraction or contracture, to the correction of deformity, or to the removal of an impediment to the action of the weaker non-contracted muscles. It is not claimed that it improves in any way the weakness of the paretic muscles. On the other hand, by mechanically lengthening the contracted muscles it takes away a certain amount of support from the limb, and while enabling it to be placed in the normal position, renders it even to a slight degree weaker than it was previously. This slight increase in weakness is more than compensated for by the advantages obtained, and if proper after-treatment be pursued is usually easily remedied. In our opinion the final results of these cases as regards benefit to the patient depend almost entirely on the faithfulness and care with which the after-treatment is carried out. Tenotomy in these cases should be regarded only as the first step in the cure, as a means to enable us to apply our other treatment, electricity and massage, at a greater advantage.

In regard to the operation itself, it suffices to say that it is on the whole a simple one, and that under proper conditions it is perfectly safe. It is too well known to need further description.

In this paper I shall deal principally with the question of the result of this operation as regards recontraction and rectification of the position of the limbs. The following table will give the results.

Spastic Paralysis.

I. E. A. B. Male, 4. This is Case I. in Dr. Bradford's article. The child was mentally deficient and had never walked. Was unable to talk. Both legs were flexed at the knees, and the feet so flexed that the heels could not be brought to the floor. The right hand and arm was rigid. The tendons Achilles were divided under ether, November 20, 1882, at the Hospital of the Good Samaritan, and the knees pulled straight, and a plaster-of-Paris bandage applied to the feet, legs, and thighs.

Entered the Children's Hospital, November 3, 1885, to see if any form of apparatus could be made which would enable him to walk. He was then unable to stand, and the legs were flexed at an angle of 30°.

April 16, 1892. His mother writes. "From the time of the operation to 1885 he seemed to improve. Had outgrown his apparatus then and had it renewed. Wore it for two years, and continued to improve so that he could walk across the room holding some one's hand." In June, 1887, " began to have a sort of spasm or convulsion, began gradually to lose strength, and in a short time was unable to bear his weight on his feet. Now has convulsions at irregular intervals, sometimes as many as four in a day." Present condition : Cannot walk. Cannot stand even with support. Ankles very weak and limber, — one of them bent over. Thighs are straight, somewhat adducted, but not as much as before operation. Knees bent, very stiff. Can touch heels to floor, but toes are inclined to tip up when heels are on floor. General health varies.

II. Female, 12 years of age (Case II. in Dr. Bradford's series). Has always had difficulty in walking. When doing so was unable to place her heels upon the ground, and the knees turned badly. Intelligence not impaired. In the spring of 1884, Dr. Bradford performed tenotomy of both Achilles tendons. Six months later, the patient walked entirely upon the flat of the foot, striking the heel normally at every step, and there was no tendency to relapse.

April 26, 1892. Spastic paraplegia. Thighs adducted, knees slightly flexed, hamstrings tense when at rest. Can abduct thighs well. Toes drag a little, but heels touch ground well. Able to walk some miles without fatigue.

III. Female, 5. Imbecile with spastic paraplegia. Operation by Dr. Bradford. Tenotomy of both Achilles tendons October 8, 1885. Discharged October 28, 1885. No further account obtainable.

IV. Male, 5. Spastic paraplegia, operation by Dr. Bradford, October 4, 1886. Tenotomy of both tendons Achilles; knees straightened and included in stiff bandage. April 26, 1892. Letter from patient's father. Walks with crutches. Hips seem to be locked together. Tendency of knees to cross. Touches his heels to floor "too much," but stiffly. Knees bent, but can be slowly straightened. Right ankle bends over with the least weight; left ankle stronger; upper extremities unaffected.

V. Male, 11. Spastic paraplegia. Operations by Dr. Bradford, December 15, 1886. Tendons of left hamstrings cut by open incision. January 2, 1887. Tendons of internal hamstrings and of biceps and fasciae divided in

right leg, and leg forcibly straightened.   Discharged March
2, 1887.

Letter from Superintendent State Primary School, May
3, 1892, says: "Can walk, but not easily.   Legs are
crooked considerably at the knee.   Joints are stiff, and
he walks on one side of right foot.   In my opinion he
derived but little benefit from the operation."

VI. Female, 11.   Feeble-minded.   Spastic hemiplegia
of left extremities and face.   Operation by Dr. A. T.
Cabot, September 2, 1887.   Subcutaneous section of
plantar fascia and left tendo Achilles.   Discharged,
improved, September 21, 1887.   No further history
obtainable.

VII. Male, 9.   Spastic paraplegia.   On admission,
April 19, 1888, unable to stand without a cane.   Rests
weight mostly on the right foot, as there is a marked
talipes equinus on the left, and contraction of the left
hamstring muscles.   The left foot is rigid.   There is
slight talipes equinus of the right foot, which can be
brought into fair position.   Operations by Dr. Bradford,
May 4, 1888.   Subcutaneous tenotomy of both Achilles
tendons and of the insertions of the adductor longus.
May 19, open incision and section of hamstring tendons of both knees.   June 27, discharged relieved.

Letter from father, April 27, 1892.   No adduction of
thighs.   Can touch heels to the ground.   Can stand
with his legs straight, but usually flexes them a little
at the knees.   His gait is lumbering and at times rolling.   There is a tendency to swing the left leg inwards
and to toe in; this is less on the right.   Puts his feet
down with a slap, and when tired drags his feet.   "The
operation had the effect of giving him more freedom
of motion at the knee-joint."

VIII. Female, 4. Spastic diplegia. On admission to hospital, September 24, 1888, unable to stand alone. Slight contraction of both Achilles tendons. September 25, operation by Dr. A. T. Cabot, subcutaneous tenotomy of both Achilles tendons. Discharged, relieved, October 13. March 18, 1890, walks, when helped, with a characteristic spastic gait; cannot walk unaided.

April 20, 1892. Report of a friend. Can stand alone with apparatus, but not without it. Can walk but little, and not at all unaided. Does not touch heels to the ground when walking, but can touch them. Feet are in good position.

IX. Male, 12. Spastic diplegia. Admitted to hospital September 29, 1888. Then walked with short steps and with difficulty, dragging the feet; the knees "scarcely flexed" and there is a slight tendency to walk on the toes. Ankle clonus. October 8, operation by Dr. Bradford. Tenotomy of both Achilles tendons, of hamstrings in the popliteal spaces, and of the insertions of the adductor longus and gracilis (on each side?). Seventeen days after operation much improved. No further information obtainable.

X. Male, 11. Spastic diplegia. Admitted to hospital October 13, 1888. Unable to stand without crutches, and then only for a short time. Marked contraction of adductors and hamstrings. October 15, operation by Dr. Bradford, subcutaneous tenotomy of tendo Achilles, hamstrings, and adductors in both legs. April 21, 1889, operation by Dr. Bradford; division of the belly of the tensor vaginae femoris and fascia lata on the left, division of the adductors and of the hamstrings, all by open incision.

April, 1892. Letter from father. Walks, but only with crutches. Slight adduction of thighs. Touches heels to the floor. Legs rather stiff at the joints. Legs apparently straight when lying down.

XI. Female, 2. Spastic hemiplegia, July 22, 1889. Has never walked, stood, or crept; cannot sit alone. December 23, 1889. Sits alone. Cannot stand alone, when supported touches only toes. Operation January 4, 1890, in presence of Dr. Bradford, tenotomy of both Achilles tendons. Patient has reported constantly since. She has had convulsions not infrequently. There is much weakness of the lower extremities, but little tendency to recontraction. No mental impairment.

This patient was operated upon a second time, in November, 1892, by Dr. Burrell, who cut the hamstring tendons under both knees by open incision, and also again divided both Achilles tendons subcutaneously. The amount of contraction then present at the ankles was variously recorded by different observers, but it seems certain that it was less than previous to the first operation. The patient was last seen by Dr. J. L. Morse, August 23, 1893, and the condition of the limbs was excellent as regards recontraction. The mother states that there has been a decided improvement in the condition of the lower extremities after each operation, and would be glad to have another operation to relieve the contractures of the right upper extremity, which has never been attempted. She has walked a couple of blocks (probably with some one's hand) several times. Both ankles allow flexion beyond a right angle, and the knees can be completely extended.

XII. Female, 13. Spastic diplegia. Admitted to hospital February 17, 1890. Able to stand only with support. Lower extremities adducted, flexed; knees held at an angle of 120°; hamstrings tense. Marked flat feet. February 27; operation by Dr. Bradford. Tenotomy of both Achilles tendons. Division of hamstring tendons and of tendon of adductor longus in both legs.

March 14. Walks daily without assistance, but on toes, heels not touching floor. April 5, discharged relieved.

April 27, 1892. Seen by Dr. Morse. Mentally backward. Both upper extremities rigid. When sitting, legs are held abducted, knees flexed about 30°, feet everted, ankles at right angles. Can straighten knees when she lies down. Can stand and walk a few steps with crutches. In walking the body is bent forward, thighs flexed on the body, knees flexed, feet everted and flat. As a rule, touches heels. Tendency to abduction of thighs. Ankles can be flexed (dorsally) to more than right angles.

XIII. Male, 12. Spastic paraplegia. Lower extremities in condition of spastic contraction at hip, knee, and ankle on May 1, 1890, and operation advised. May 5, operation by Dr. Lovett. Tenotomy of right tendo Achilles. Division of internal hamstrings by open incision and of biceps subcutaneously on the right. June 6, tenotomy of left tendo Achilles, and division of hamstrings and tendon of adductor longus in left lower extremity. June 25, discharged improved.

XIV. Female, 8. Spastic paraplegia. June 18, 1890. Walks on tiptoe, with knees bent and thighs slightly adducted. June 30, operation by Dr. Bradford, tenotomy of tendo Achilles and division of hamstrings in both limbs. July 12, discharged relieved.

April 21, 1892. Letter from father. " Can walk, but makes very hard work of it. . . . When she puts the foot down, she lets it down heavy." Wore apparatus for some months, but then left it off. Knees bend in badly, and strike when she walks. Can put heels on the floor which " is the only change I can see," as result of the operation.

XV. Female, 9. Weak-minded. Subject to convulsions. Right spastic hemiplegia of extremities, face not involved. Admitted September 9, 1890. Decided atrophy, and much spastic contraction of right upper extremity; shoulder almost immovable; elbow, which is held flexed, can be straightened passively; hand drawn strongly to the ulnar side, with partial dislocation of wrist. Thumb strongly adducted and flexed in the palm of the hand, and passive straightening is painful. September 15, operation by Dr. Burrell. Division of tendon of flexor brevis pollicis and of palmar fascia at base of thumb, also of tendon of flexor carpi ulnaris at wrist; open incisions. October 10, discharged relieved.

April 18, 1892. Mother states that the thumb remains extended, but that there is little or no power in it. Hand still abducted.

This operation, which was strongly urged by me at the time, was, so far as I know, the earliest one performed on the upper extremity in this hospital. The patient unfortunately did not return for electrical treatment, and hence the paralysis or weakness of the thumb was not improved, although its position was rectified.

XVI. Male, 4. Spastic paraplegia. Mind clear. Admitted to hospital January 12, 1891. Feet in equinus position, cannot be dorsally flexed beyond right angles. January 17, operation by Dr. Burrell. Tenotomy of both Achilles tendons, right subcutaneously, left by open incision. February 3, discharged relieved. March 10, 1891, both ankles now movable and child walks well with support. April 27, 1892, mother thinks he was much benefited by the operation. Gait much better than before operation. Touches his heels first and firmly. Wears out shoes in front on the outside.

XVII. Male, 11. Spastic paraplegia. Admitted to hospital April 23, 1891. Hamstrings tense and knees flexed in walking, heels raised from floor, double equinus. April 25, operation by Dr. Bradford. Subcutaneous tenotomy of hamstrings and Achilles tendons on both sides. June 17, walks with considerable agility, can run awkwardly. Hamstrings apparently not contracting. Discharged relieved. June 22, Dr. Bullard states that there is still slight flexion at left knee; otherwise lower extremities can be extended perfectly when lying down. Passive motion, except knee as above, everywhere good. When standing holds body flexed on thighs, and flexes both knees, this position being due to weakness and not to contracture. Unsteady on feet and afraid of falling.

April, 1892. Can walk alone but prefers to have support. Gait stiff. In walking, thighs held flexed at about 15°, and knees at about 10°; touches heels to ground. Feet inverted, and he walks on the inner borders, especially on the left. Hamstrings slightly stiff, easily corrected. Ankles rather flaccid, motion perfectly free.

XVIII. Female, 4. Weak-minded. Convulsions. Spastic paraplegia severe, with extreme adduction of thighs. Hamstrings somewhat contracted. Double equinus; stands on toes. Admitted July 30, 1891. Operations by Dr. Cushing. July 30, subcutaneous tenotomy of adductor longus on each side. August 18, subcutaneous tenotomy of both Achilles tendons. September 11, discharged relieved. April 27, 1892, seen for first time since leaving hospital. Cannot walk nor stand alone. When standing with support, body flexed on thighs, and knees flexed, feet in valgus position. No strength in lower extremities; when lying down, right thigh nearly straight, right knee very slightly flexed, right foot and ankle very loose and tendency to valgus; left lower extremity in similar condition, but more contraction, slight, of adductors of thigh, and moderate amount in hamstrings.

XIX. Male, 5. Severe case of spastic paraplegia. Admitted to hospital August 19, 1892. At that time could not stand alone. Thighs adducted, knees flexed, and heels do not touch the floor.

Operation by Dr. Burrell. Division of hamstring tendons in both legs by open incision, and of both Achilles tendons subcutaneously October 6. Discharged relieved. In November, 1892, this patient was brought to the hospital, and there was little or no recontraction.

RÉSUMÉ.

I.	M.	Dr. Bradford.	Achilles tendons.	No recontraction.
II.	F. 12.	" "	" "	" "
III.	F. 5.	" "	" "	
IV.	M. 5.	" "	" "	" "
V.	M. 11.	" "	Hamstrings.	Recontraction ; amount unknown.
VI.	F. 11.	Dr. Cabot.	L. Achilles.	
VII.	M. 9.	Dr. Bradford.	Achilles and hamstrings both sides.	No recontraction.
VIII.	F. 4.	Dr. Cabot.	Achilles tendons.	Slight contraction.
IX.	M. 12.	Dr. Bradford.	Achilles, hamstrings, adductors of thighs.	Much improved in 17 days; no further record.
X.	M. 11.	" "	Achilles, hamstrings, adductors.	No recontraction.
XI.	F. 2.	" "	Achilles.	Some recontraction probable.
XII.	F. 13.	" "	Achilles, hamstrings, adductors.	No recontraction.
XIII.	M. 12.	Dr. Lovett.	Achilles, hamstrings, L. adductors.	Achilles did not recontract.
XIV.	F. 8.	Dr. Bradford.	Achilles, hamstrings.	Recontraction of hamstrings.
XV.	F.	Dr. Burrell.	Flexor brevis pollicis and fl. carpi ulnaris.	No recontraction of thumb.
XVI.	M. 4.	" "	Achilles tendons.	No recontraction.
XVII.	M. 11.	Dr. Bradford.	Achilles and hamstrings both limbs.	" "
XVIII.	F. 4.	Dr. Cushing.	Achilles and adductors of thighs.	" "
XIX.	M. 5.	Dr. Burrell.	Achilles and hamstrings both limbs.	No recontraction, but time too short to be decisive.

As the results of these operations we find: —

I. That there is but little tendency to recontraction after division of the tendons in cases of cerebral spastic paralysis, although in the severer cases this may occur to a certain extent. There is apparently a stronger tendency to recontraction in the hamstrings (probably because they are less completely divided) than after tenotomy of the Achilles tendons. Indeed, after the latter operation we sometimes find that the foot is too lax, and this result must be guarded against.

II. There is no gain in the strength of the paralyzed limb from this operation. On the other hand, the support of the contracted muscles being removed, the limb is to that extent weaker than before.

III. The benefit to be derived from this operation consists in the correction of deformity, and more especially in the placing of the limb in a condition in which other means of overcoming the paralysis (massage and electricity) can be applied to better advantage and with more chance of benefit.

This operation is to be considered as the first step in the treatment of the severer forms of cerebral spastic paralysis in children, but only as the first step. It needs to be followed by other measures in order that the full benefit of its result should be obtained.

I desire here to express my thanks to Dr. John L. Morse for his assistance in obtaining the histories of these patients.

WHAT IS A NORMAL SPINE.

BY HAYWARD W. CUSHING, M.D.

THE medical practitioner is often called upon to examine patients for spinal lesions. He uses his eyes. He uses his fingers. Sight and touch are the senses on which he must depend for his objective examination. The true significance of the data thus obtained is at times a question of great importance. In many cases, it is true, he has no difficulty in satisfying himself that a pathological process is, or is not, present; but occasionally, and really oftener than one would suppose, he finds variations from the usually described normal type, which if not taken at their true worth may cause him to err in his diagnosis in either direction. A normal back may be considered the seat of tubercular disease ; or an affected spine may escape detection with disastrous results. I will cite three cases.

1st. Boy, two years old; fat, rachitic, and had never walked. Was brought for examination on account of a prominence of the back which had been sufficiently conspicuous to attract attention. The child sat in the usual rachitic attitude, with a long gradual spinal curve; but there was also present an additional irregularity and prominence of the eleventh dorsal spine. Irregular muscular spasm was seen when the spine was hyper-extended, and, if also watched for some time while sitting, an inclination to support the trunk with the extended arms. Is this commencing caries? No.

2nd. A male adult. Injury to back four years ago in a railway accident. Confined to bed for three years, during which time has had three attacks of spinal meningitis, — severe. There is great pain on motion, referred to the third and fifth dorsal spine, also great tenderness at the same point. The tip of the third dorsal spine is one-fourth of an inch to the left of the median line. Slight relief from medicinal or mechanical treatment. Is bedridden. A diagnosis here is of great importance, both for financial reasons and with reference to operation. Of what diagnostic significance is this displaced spinous process?

3d. A small child. Aet. five. Well developed. History of a fall one week previous to examination. Previous history shows that this patient's general condition has not been especially good. Brought for examination on account of occasional dorsal pain on motion, and rigidity. Examination reported negative except a slight prominence of the seventh and ninth dorsal spines and a very slight irregular muscular spasm. Back considered normal. Diagnosis, muscular strain and contusion. Seen next at end of six months, with a marked kyphosis and psoas abscess.

These cases illustrate, I think, what is stated above, that at times one sees spines which do not conform to the usual type. They may be simply a normal variation which may simulate pathological processes. In some cases they are the changes due to actual disease. Therefore the questions arise: What are these normal variations? How often do they occur? Can they be distinguished from those due to disease? In order to answer these queries I have examined many living spines with especial attention to these points. I have notes in detail of thirty especial cases in children. I have examined thirty skeletons. I have studied all the backs of statuary accessible in art museums and in private. Also numerous photographs, and anatomical plates. I have tried to find out what the spine really is anatomically. What the ideal spine is as represented by ancient and modern art; and how far the clinical spine, as ordinarily met in health and disease, conforms to this type. I will first describe the spine anatomically as I have found it in the skeleton, and state the details which attracted my attention.

CERVICAL SPINE.

Its posterior profile. Often as shown in standard anatomies. The spines of the axis; the 6th and 7th vertebrae most prominent; a depression between, whose greatest depth is opposite the 3d, 4th, and 5th. This inequality in length allows for extension of the head. The depth and outline of this profile curve also varies with the position of the head, being shallowest when the head is flexed, and deepest during extension. But the profile outline showed considerable variation, a knowledge of which is of value where fractures or displacements are suspected. These variations are caused by a difference in the comparative length of the cervical spines, especially of the atlas and axis, and of the 5th, 6th, and 7th. The atlas, which is usually drawn as short, and not to be felt in the living subject, was found to be nearly as long as the axis eleven times in twenty-five spines. There was a variation in the comparative length of the 2d, 6th, and 7th of from a quarter to one inch. The 6th and 7th varied from nearly equal length to the 6th being an inch shorter. The 6th was usually as much longer than the 5th as the 7th exceeded the 6th in length.

The anterior surface of the atlas has a bony projection in the median line which can be felt through the mouth. This is sometimes quite distinct, and might simulate an exostosis or an anterior displacement. It is the point of attachment for the longus colli muscle. The transverse processes are much more prominent laterally than those of the other cervical vertebrae. It is below and anterior to the tip of the mastoid, and a little nearer the median line, and can be indistinctly felt at this point.

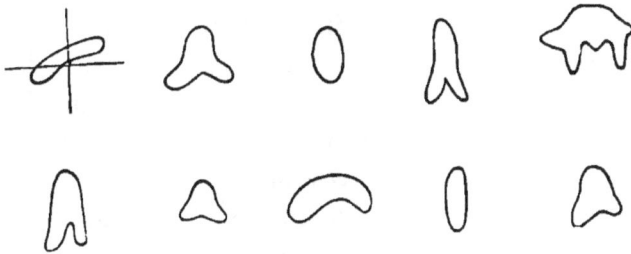

OUTLINE OF THE TIPS OF THE SPINE OF THE AXIS.

The conspicuous feature of the axis which interests us in this connection is its spine. Its length has been mentioned. It is the first bony point usually felt in the median line below the occiput, the spines of the 6th and 7th vertebrae being generally the next. It is also a fact that the outline of its tip is very irregular. The accompanying diagrams illustrate this. They are drawn from actual specimens.

The 3d, 4th, and 5th cervical spines do not attract especial attention, except that occasionally they can be felt. Their tips are generally bifid. The 6th and 7th form the lower boundary of the cervical depression. When long, the edge of the depression is abrupt, and in

some cases might simulate anterior displacement. No
other peculiarities were noticed other than those usually
described.

The Dorsal Spine.

Here, in addition to those characteristics usually
described, the contour and arrangement of the spinous
processes first attracts attention. The profile curve is
noticed to be irregular. This is found to be caused by
differences in length, direction, and contour of the spinous
processes. This irregularity in profile formed depressions,
the deepest of which in twenty-five spines was from one-
half to five-eighths of an inch. It was most frequently
seen over the 3d to 6th dorsal vertebrae. The 7th cervical and 1st, 2d, 3d dorsal spines have a much less
inclination than those next below. The spines therefore can be felt and seen as well as the interval between
their tips. From the 4th to 10th, inclusive, the direction
changes abruptly, the spine of the 4th forming quite
an acute angle with that of the 3d. From the 4th to
the 11th, the slant is such that the spines, although
perhaps longer than the upper three, are so closely
shingled upon each other from above downwards as to
form with the ligaments an almost flat surface. Below
the 10th, the spines become more horizontal, and again
show an interval between their tips. The prominent
spinous tips are therefore the 7th cervical, 1st, 2d, 3d,
10th, 11th, 12th dorsal, the 1st to 5th lumbar, inclusive.

The contour of the dorsal spines is quite varied, as the
accompanying figures show.

The relation of the transverse to the spinous processes
is a factor in the surface contour of the dorsal region.
Here the transverse processes not only arise posterior to

the articular, but also from the 4th and 5th begin to bend backward so that lower down their backward projection equals, or in some cases slightly exceeds, that of the spinous processes. The maximum of this backward curve

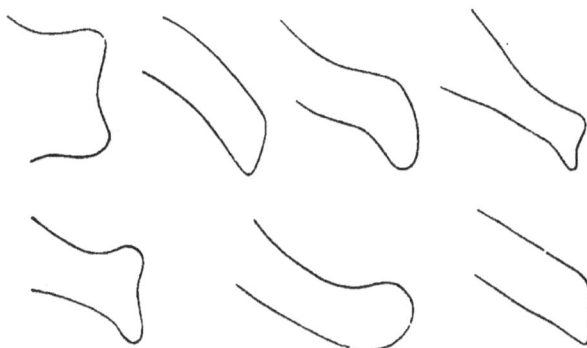

CONTOUR OF SPINOUS PROCESSES OF DORSAL VERTEBRAE.

is opposite the 7th and 8th spines; it then gradually diminishes to the 11th. Attention may also be called to the plane of the articular surfaces, which does not permit antero-posterior motion.

The spines were also noticed to be distorted to a greater or less amount in their backward direction. This was found to cause lateral deviation from the median line of the spine. This was most frequent in the dorsal vertebrae (3d to 9th and 10th).

4th Dorsal.
5th "
6th "
7th "
8th "
9th "
10th "
11th "
12th "

LATERAL DEVIATION OF SPINOUS PRO-CESSES OF DORSAL VERTEBRAE.

The deviation was sometimes to the right, sometimes to the left, sometimes to the right and left in the same spine. It equalled in some cases one-quarter to one-half of an inch, and included one or several vertebrae. The importance of these variations in diagnosis of injuries of the spine is self-evident.

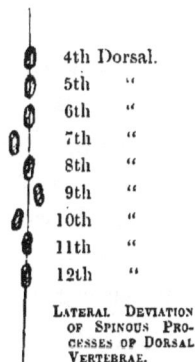

LUMBAR SPINE.

This was found to be more uniform as regards its curves, profiles, and contour in detail than any other region investigated. The depression at the lumbo-sacral articulation is a valuable landmark, but is not always distinct.

All the preceding describes the skeleton. But we must remember that the living spine is covered with ligaments, muscles, fat, and skin. These tissues change and modify the outlines of the skeleton, in fat or well-nourished persons to a marked degree. The human back shows certain curves, rounded surfaces, depressions, and grooves. As a rule the bony prominences of the skeleton are at the bottom of these valleys, since the depressed portions are filled by the bellies of the muscles which form the surface elevations. In disease these structures waste, and the true skeleton outline comes more and more to view. The structures which produce the main curves of the back are the muscles. They form a vertical sulcus extending from the occiput to sacral region. As above stated, this dorsal groove varies with the muscular development of the subject, and the amount of fat present. Also with the position of the back. In men (less so in women and children) it is usually more conspicuous in the cervical, low dorsal, (10th, 11th), and lumbar region. In the erect attitude, it is as described deepest in the cervical and lumbar regions, and flattened over the dorsal. The reason of this flatness between the 4th to the 10th dorsal spines, sometimes including four vertebrae, sometimes six, and sometimes eight, has been indicated above. It is caused by the backward curve of the transverse processes at these points and the extreme slanting of the spines themselves.

The prominent spines are the 7th cervical, 1st to 3d dorsal, 11th, 12th dorsal, and the lumbar. But this varies sometimes, and backs are occasionally seen (one in twenty-five) with almost the same length and slant of the spinous processes in the dorsal region. In such a back, more tips would be visible in the vertical sulcus.

In a stooping position, the groove disappears below and the spinous tips become more prominent. The cervical groove is deeper. Any irregularity in the length or direction of the bony spines is also seen here, and is the cause of the projections at individual points which might be thought to indicate a beginning kyphosis. They are only spines which happen to be a little longer or have a different inclination from those immediately adjacent.

When prone, the cervical groove disappears while the lumbar sulcus extends upward, sometimes to the mid-dorsal region.

The study of the back from statuary, at least from plaster casts, is, as a rule, unsatisfactory. The dorsal groove is often shown as an unbroken sulcus, perhaps flattened over the 7th cervical to 3d dorsal spines, and no spinous processes are indicated. One exception is the Capitoline statue of the "Spinario," the figure of a boy seated with one leg crossed in the effort to extract a thorn from the sole of the foot. This is a stooped spine. It shows a depression at the 4th, 5th, and 6th dorsal spines, also a wavy median line. The "Dying Gaul" is another figure which shows a back corresponding with my observations. Female backs are from models with so much subcutaneous fat that bony prominences are obscured and the dorsal groove quite shallow. I have studied few statues of children. These were mostly of the Raphael cherub type, which for the present purpose arc of little value.

These investigations show, that the clinical, and for us the practical, results correspond fairly well with the anatomical, and show the same variations that the latter would lead one to expect; that the artistic spine does not; that in examining patients it is important to consider : —

Position or attitude, — erect, stooping, or prone ;

Age and sex ;

Size of muscles ;

Amount of fat.

And now the results. What conclusions can be drawn from this mass of variations?

That within certain limits no two spines are alike any more than individual faces.

Spinal lesions, aside from subjective symptoms, are alleged to present : —

Muscular spasm causing limitation of motion to a greater or less extent;

Irregularities of spinous processes either from the median, or normal profile line, — i. e. kyphosis or irregular scoliosis.

The above results show that we find : —

Slight or distinct irregularities of spinous processes with reference to length ;

A vertical sulcus more or less varied ;

Slight kyphotic projections of single spines ;

A marked depression above the 7th cervical ;

Lateral deviation of spines even to ½ and ⅝ths of an inch ;

A bony process, at times almost an exostosis, on the anterior surface of the atlas.

These conditions, therefore, when found, cannot be considered as indicative of injury or disease *per se*, and other symptoms must exist to permit such a diagnosis.

Muscular spasm. — This must be examined for with equal care. It is uncertain and variable. Fright, involuntary acts, pain, *e. g.* due to muscular twists or strains, all cause a condition which simulates the spasm due to disease. Hence, with such a spine as above described, a false diagnosis might be easily made, — a rachitic curve treated as caries. Such cases can only be accurately recognized by frequent observations.

HIGH TEMPERATURE IN CHRONIC JOINT DISEASE.

BY ROBERT W. LOVETT, M. D.

A PRELIMINARY report calling attention to the presence of high temperature in the early stages of chronic joint disease was presented in 1890[1] by the author. The present paper is the result of the continuance of those observations, which have seemed to establish those conclusions, and to carry the matter further. The conclusions about to be presented rest upon a uniform series of observations, which now amount to 1050 in number. They have all been taken under the supervision of the writer for the most part at the surgical out department of the Children's Hospital.

The investigation has covered a period of four years, although not continuously, and has continued during all seasons of the year. From July 1, 1890, to July 1, 1891, for example, 627 observations were made in hip and spinal disease. It seems hardly worth while to attempt a tabular presentation of these figures, but rather to show the conclusions which they warrant.

The question at issue is, do children with tubercular joint disease suffer from enough systemic infection to give rise to a hectic condition? If so, what is its significance?

---

[1] Boston Med. and Surg. Journal, April 17, 1890.

The observations to be noted were made on all cases of hip disease and Pott's disease coming to the clinic, and not on selected cases. The thermometer used was sent to the Yale observatory and standardized. The hour of observation was between 3 and 5 P.M. and the mouth temperature was taken in most cases, but in very young children the rectal temperature was necessarily substituted for this. The conditions under which the observations were made were therefore practically constant. As a check to these observations, the temperatures of healthy children coming with their brothers and sisters to the same clinic and under the same conditions were taken. They were as follows.

98.3	99.1	99.3	98.	99.3
98.5	98.1	98.	99.3	98.5
97.3	97.3	99.1	99.3	97.3

As a matter of contrast, it is of interest to look at a series of hip cases in December, 1890, taken consecutively from the writer's record book. They were as follows : —

99.8	100.	99.	101.2	101.4
99.	99.8	98.6	100.	100.
100.2	100.	101.6	100.4	99.

From the same period are temperatures of the following cases of Pott's disease : —

99.8	101.	101.2
100.2	98.6 [1]	100.
103.	99.8	99.6

Similar figures to these are shown by the records wherever one opens them, and the number of observations is surely enough to enable one to generalize without presenting in detail the hundreds of figures which are under consideration.

[1] Cervical disease without abscess which has completely recovered.

High afternoon temperature is a constant accompaniment of disease of the spine, hip, and knee undergoing ambulatory treatment. The average temperature is higher in Pott's disease than in hip disease, and the rise is least marked in knee-joint disease.

A rise of temperature to at least 99° is present in the great majority (90%) of all cases, whether acute or chronic, slight or severe. It rises to 103° or 104° in severe cases, without necessarily being associated with pus formation.

THE CONDITIONS UNDER WHICH HIGH TEMPERATURE IS PRESENT.

The temperature is lowest in cases which have little swelling about the joint and in which acute symptoms are absent. In this class of cases the temperature is often normal. In acute cases, where there is excessive joint tenderness and thickening, the afternoon temperature most often rises to 100° or 101°, and becomes lower with the subsidence of the acute symptoms. The presence of an unopened abscess does not necessarily give rise to higher temperature than would exist in cases of like severity without the presence of pus. When an abscess is forming the temperature is generally high because of the severity of the case. The highest temperature of all is found in late cases of hip disease, and especially Pott's disease, where rapid disintegration is taking place through sinuses which discharge profusely. Under these circumstances an evening temperature of 103–104° is not uncommon. Cases where pulmonary tuberculosis is present have of course been excluded from these observations. It may be said in general that patients who have discharging sinuses almost invariably show a very decided rise of temperature.

The most interesting and the most significant fact of all is that the high temperature is not noted during treatment by recumbency in any but the severest cases. In very many of the cases observed where there had been a continuous evening temperature of 100° or more while under ambulatory treatment, it has been observed that confinement to bed in the wards has resulted in a normal evening temperature within one or two days. This statement rests upon a fairly large number of observations.

No one believes in judicious ambulatory treatment more strongly than the writer; but it seems to him that these observations offer us a hint to restrict the activity of these tuberculous children. If the writer is correct in his conclusions, this high temperature at night represents absorption from the focus of disease. Absorption is notoriously more active when the circulation is stimulated by too much exercise. Which of us with a septic wound of the hand and an evening temperature above the normal would not keep as quiet as he could in the hope of preventing further systemic infection?

It seems to the writer that we have in this matter of temperature a guide which need not be neglected. Restricting the activity by some hours of enforced rest has again and again been demonstrated in this series of cases to lower the temperature. This, it is easy to see, means less rapid tissue waste, it means theoretically less danger of tuberculous generalization, and it means practically in many cases marked and rapid increase of appetite, gain of flesh, and improvement in the general condition, especially in the severer and more acute cases.

Whatever apparatus is being used, however carefully the joint is protected, and whatever may be the merits of restricted activity on other ground, the writer begs to call especial attention to the subject of high temperature as a guide.

The Diagnostic Value of High Temperature.

In the preliminary report it was asserted that high temperature was likely to be a useful physical sign in diagnosticating tuberculous joint disease.

Again and again it has served the writer in this way, although it is only of value in connection with other symptoms. The observations have shown very clearly that elevation of temperature is present in connection with the earliest symptoms of hip disease, for example, so that it exists early enough to be of use in beginning cases. It is valuable as a diagnostic sign in those cases e. g. of a slightly stiff and painful hip, where there may be no history of a fall, and where the signs are not sufficient to warrant a positive diagnosis of hip disease. The presence of an evening temperature of 100° under these circumstances greatly increases the probability of tuberculous disease. In the same way obscure affections of the spine, such as slight stiffness with perhaps some little lateral deviation of the column, may be considered more or less suspicious according to the presence or absence of high evening temperature. There is no need of further delay in making special applications of so obvious a matter.

The Prognostic Value of High Temperature.

If a case of hip, joint, or spinal disease shows a persistent high evening temperature (100–102°) in the early stages of the disease, the prognosis is in the writer's opinion a bad one, as in these observations such cases have proved to be very painful and severe ones, or they have been of that malignant and rapid variety which leads to destructive disintegration of the joint.

If the temperature, on the other hand, is a low one .(99–100°) in the early stages of the disease, the case is more likely to be either mild and short, or very long and of the painless type.

Cases under treatment soon settle down to a regular range of temperature, perhaps 99.5° to 100°, and this temperature remains constant so long as the conditions remain the same. A rise in that temperature is a sign of coming mischief in most instances, betokening either an acute, painful attack or some joint mischief. A very high temperature (103°) has in several instances under the writer's observation been the first sign of tubercular meningitis.

Convalescent cases under favorable progress should have a normal temperature; a rise of temperature has in several instances been the sign of coming relapse.

In short, in any case, a rise above the accustomed temperature of that case betokens coming or present mischief, and is a sign worthy of attention.

Such are the facts and conclusions to be derived from these 1050 observations. It is impossible to do more than state the facts and their obvious application, trusting that they may appeal to some who may find them useful.

To the writer's mind the chief significance of them lies in the fact that they demonstrate the existence of a general as well as a local infection, and give another and satisfactory reason for restraining the activity of these tuberculous children.

TREATMENT OF INFANTILE CLUB–FOOT.

BY E. G. BRACKETT, M. D.

IT is recognized that in infantile club-foot the contraction of the soft parts offers a greater obstacle to the correction of the deformity than does the osseous malformation ; hence the object of all methods of treatment is, first, to so overcome this resistance that the foot can be brought to a position of over-correction without force ; second, to hold it in this position until, during growth, such adaptive changes occur that the deformity shows no tendency to return even when the foot is in active use, as in walking. The establishing of the accurate balance of the antagonism of muscles and other soft parts, when the foot is in normal position, constitutes the important element in the permanency of the correction. After the first resistance has been overcome, it is often found possible to easily bring the foot to a position of over-correction, but when left to itself, influenced only by gravity, it returns in a greater or less degree to its old position of deformity, and this is especially so when the muscles are in action. Unless the after treatment of retention in over-correction is persisted in until the child is able, without special effort, to walk with the foot in a natural position, a relapse will follow. It is therefore necessary, whatever means is employed, that all methods should have thorough persistency, without relaxation even for brief intervals, as one object is to shorten the elongated structures; and tension on these, which by the treatment are given the opportunity to contract, should not be allowed more than is needful for the change of the bandages or apparatus. This is true not only of the methods which aim at gradual restoration, but still more in those which aim at the rapid correction, as in the early operations here to be described.

In order that the treatment should be thorough, and the result permanent, the most important object should be to bring about a normal relation of all parts, through growth, rather than to correct an abnormal position. Although in these small feet the condition of the contracted soft parts is the most important factor in the prevention of rectification; the contraction of the yielding structures is only one element to be considered so far as permanency of a perfect result is concerned; the elongated condition of the antagonistic parts must be taken into a thorough consideration, otherwise, after a complete correction of a distorted foot, there are no structures which will hold it in a normal position.

The object of the treatment then is to bring about an equalization of all the structures, and to accomplish this, retention in the corrected position must be persisted in until this change becomes complete; that is, the shortened structures must be elongated until they no longer form an obstacle to complete correction, and also their antagonists must be contracted so that they serve as an obstacle to the return of the foot to its original deformed position. To accomplish this, the foot must not only be over-corrected, but must be so held long enough for this change to take place, until the foot, when at rest, and influenced only by gravity or when put into active use, as in walking, shows no tendency to assume its distorted position.

Distinction must be made between these active and passive conditions. If a corrected foot when in use is drawn strongly toward the direction of its old malposition, and is left in this condition without firm retentive appliance, it is gradually turned until the old distortion is nearly as great as at first. If the free use of the foot is allowed before the condition of the soft parts is such that a return is not easy, this use strongly tends to the return of the distortion, and unless the foot is held firmly, is greater than the corrective force of the appliance, and relapse may therefore occur, even though the care of the foot is not relaxed. The only test of value in determining when apparatus shall be discontinued, is the position of the foot when left to itself, during the act of walking. So long as there is any tendency to the return of the deformity, or until there is decided resistance to a force applied in this direction, it is not safe to omit retentive apparatus.

The most common defects of imperfectly corrected feet, or of feet which have been corrected but have relapsed, are first, the inversion of the anterior portion of the foot, the toes; second, the depression of the outer border of the foot, as seen by the resistance to elevation of the cuboid; and third, limitation of extension of the ankle joint. The first condition is the most persistent, and is frequently seen when the foot is in other respects perfect, but the flexors of the toes and adductors of the foot have not been sufficiently elongated to allow the natural position in walking, and their long-continued action results in the gradual turning of the foot, with a return of the deformity in the same manner as occurs in the development of paralytic club-foot.   This condition is often the result of too early removal of apparatus, or of use of apparatus too short to hold the anterior part of the foot and the toes. The cause of this is obvious; although the correction has been carried to completeness, the elongated structures offer no resistance whatever to the return of the foot when left to itself, and the weak muscles are not brought into equal action with their opponents.

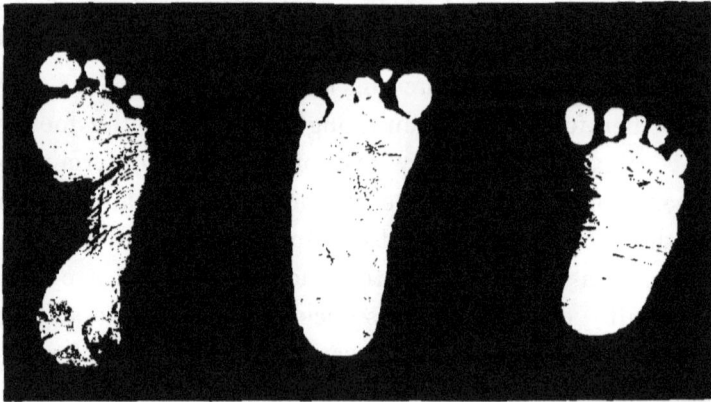

EXTENSION OF ANKLE   PERFECTLY COR-   FOOT WITH
LIMITED TO 90°.   RECTED FOOT.   ADDUCTION.

The second condition is often due to imperfect correction of the resistance, and soon appears if the adduction of the foot is allowed to persist.

This is shown in the right imprint above.   The middle imprint shows the usual condition found in a perfectly corrected foot in these young patients.   The left represents an impression of a foot in which the extension at the ankle joint was limited to 90°, and there existed still a slight contraction of the plantar fascia; the print resembles a normal adult sole, but one which in a case of this age would require careful watching for fear of relapse.

The other defect which strongly predisposes to relapse is limitation of extension in the ankle joint to or beyond a right angle, and this is true when the deformity in all other respects has been entirely corrected. If motion at the ankle joint is free, the foot is allowed to be held in any position during walking, and may be kept in a plane parallel to the axis of motion of the knee. In this way, the weight falls on the foot in the normal direction. If, however, contraction of the tendo Achilles prevents the foot from remaining flat on the ground while the knee is bent, as in the latter part of the step, the loss of this motion is compensated by adduction of the foot. In this position, in the latter part of the step, as the heel is raised from the ground, the foot is rolled to a position of inversion, and the weight is borne on its outer border. With the weight thus displaced outward on the ankle joint, each step in the act of walking exerts a force which turns the foot to the varus position. This is true both when the foot is left to itself, and when held imperfectly by apparatus.

If the important element in the treatment of these cases is the re-establishment of a normal motion of all parts of the foot, then the earlier the treatment is begun, the easier will this change be accomplished, so that at an age when the child should walk, the foot may be in a normal condition for use. And for this reason the method here advocated of immediate correction is used.

This method consists of early operation, which may be preceded by the application of plaster-of-Paris bandages for a short time, and used with the object of correcting the varus only, without reference to the equinus, or in the early operation without this preliminary treatment. In case the operation is attempted at once, it may be done any time, preferably as early as possible, waiting only for the strength and the general condition. Should the child be weak or not taking nourishment well, it is best to wait until such conditions have been corrected, and the growth is well established.

The operation consists of subcutaneous tenotomy of all the parts which obstruct the complete restoration. This in most cases consists of — division of the plantar fascia, the tendons of the tibialis anticus and posticus, the ligament of the scapho-astragaloid joint, and last, that of the tendo Achilles. After the tenotomy of the first three, the foot is further forcibly corrected by the hand, and a division at the resisting parts carried to such a point that the foot can be easily brought beyond the normal plane, after which the tenotomy of the tendo Achilles is done and the foot placed in plaster.

It is essential in the operation that while under ether, after complete division, the foot should be brought to beyond a normal position without force, so that when the plaster is applied force may not be required to hold the foot in the position which has been gained. These plaster bandages are left on for an interval of from ten days to three or four weeks. In case the restoration has not been perfect, or not without considerable force, as happens sometimes with the more resistant feet, it is well to remove the plaster at the end of ten days or two weeks, and apply the shoes (to be described), reapplying the appliance every two or three weeks. In this way, before complete consolidation has taken place, a certain amount of gain can be had. and over-correction be obtained at the end of a few weeks, when at first this was not possible.

If, however, the restoration has been complete, and without force, it is better to keep the bandages on for a longer period of time, from six weeks to two months, in order that the foot may not be disturbed from its over-

APPARATUS BEFORE AND AFTER APPLICATION OF THE BANDAGE.

corrected position. When the bandages are removed, great care should be taken that the foot is not allowed to drop from its over-corrected position, and thus make traction on the ligaments and soft parts in which contraction is desired.

The apparatus used after the removal of the plaster

bandages consists of the ordinary Taylor club-foot shoe, with an upright extending to the waist, but which is secured to the foot by means of silicate or plaster-of-Paris bandages instead of the usual straps. It is essential that the foot-piece of the apparatus be of such dimension as to hold the whole foot firmly, that is, that the tendency of the turning in of the toes, and of dropping of the cuboid should be prevented, otherwise the muscular action is not controlled, and when the apparatus is removed, serves to pull the foot to its old distortion. The heel is bound to the sole-plate by a piece of adhesive plaster extending along the back of the calf, and brought down over the bottom of the sole-plate, and the foot and ankle is then bound to the apparatus by means of a bandage.

In the application of the apparatus, the sole of the foot is first held firmly on the foot-piece of the apparatus, with the upright extending forward and diagonally across the lower leg. By the first few turns of the bandage the heel is secured to the foot-piece, and later the anterior part of the foot. The removable guard is then put in position, and a few additional turns of the bandage are then placed around the foot and the guard together.

The feet must be held by the apparatus while use is allowed, until the adapting change takes place; the pressure of the appliance must be in the direction of the normal position, and when removed its loss is not completely compensated by muscular action; it is, therefore, necessary that the foot apparatus should hold the foot in an over-corrected position.

INDEX.